Land Rover Military One-Tonne

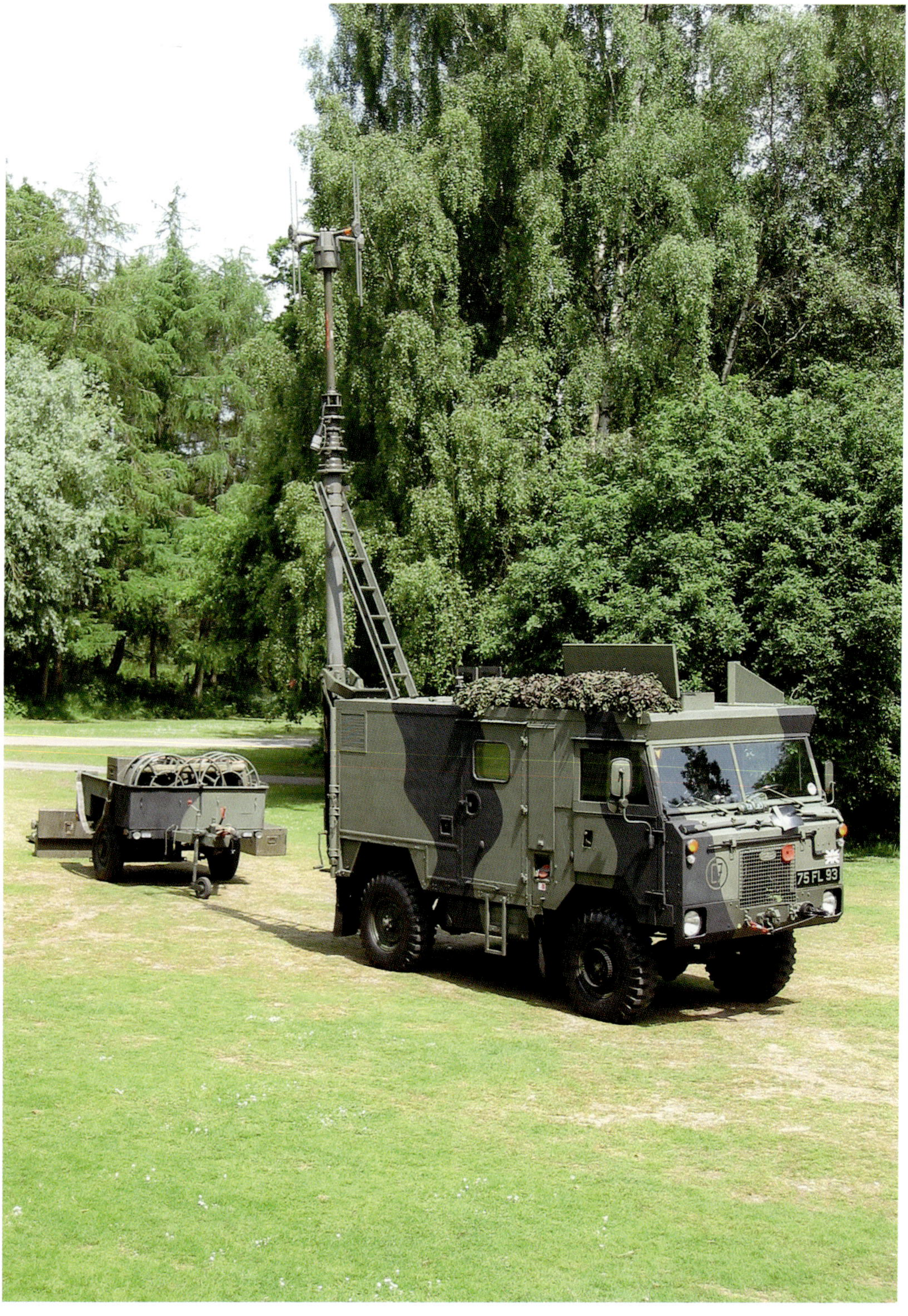
75 FL 93

Land Rover Military One-Tonne

James Taylor

THE CROWOOD PRESS

First published in 2015 by
The Crowood Press Ltd
Ramsbury, Marlborough
Wiltshire SN8 2HR

www.crowood.com

© James Taylor 2015

All rights reserved. No part of this publication may be reproduced or transmitted in any form or by any means, electronic or mechanical, including photocopy, recording, or any information storage and retrieval system, without permission in writing from the publishers.

British Library Cataloguing-in-Publication Data

A catalogue record for this book is available from the British Library.

ISBN 978 1 84797 891 2

Typeset by Shane O'Dwyer, Swindon, Wiltshire
Printed and bound in Singapore by Craft Print International Ltd

CONTENTS

	Acknowledgements	7
	Timeline	8
CHAPTER 1	ORIGINS	9
CHAPTER 2	DESIGN AND PROTOTYPES	17
CHAPTER 3	PRE-PRODUCTION AND TRIALS	37
CHAPTER 4	THE 101 IN PRODUCTION	53
CHAPTER 5	THE 101 IN BRITISH MILITARY SERVICE	73
CHAPTER 6	BRITISH 101 VARIANTS	91
CHAPTER 7	THE 101 IN OVERSEAS SERVICE	121
CHAPTER 8	AFTERLIFE	135
	Appendix: Powered-Axle Trailers	149
	Index	158

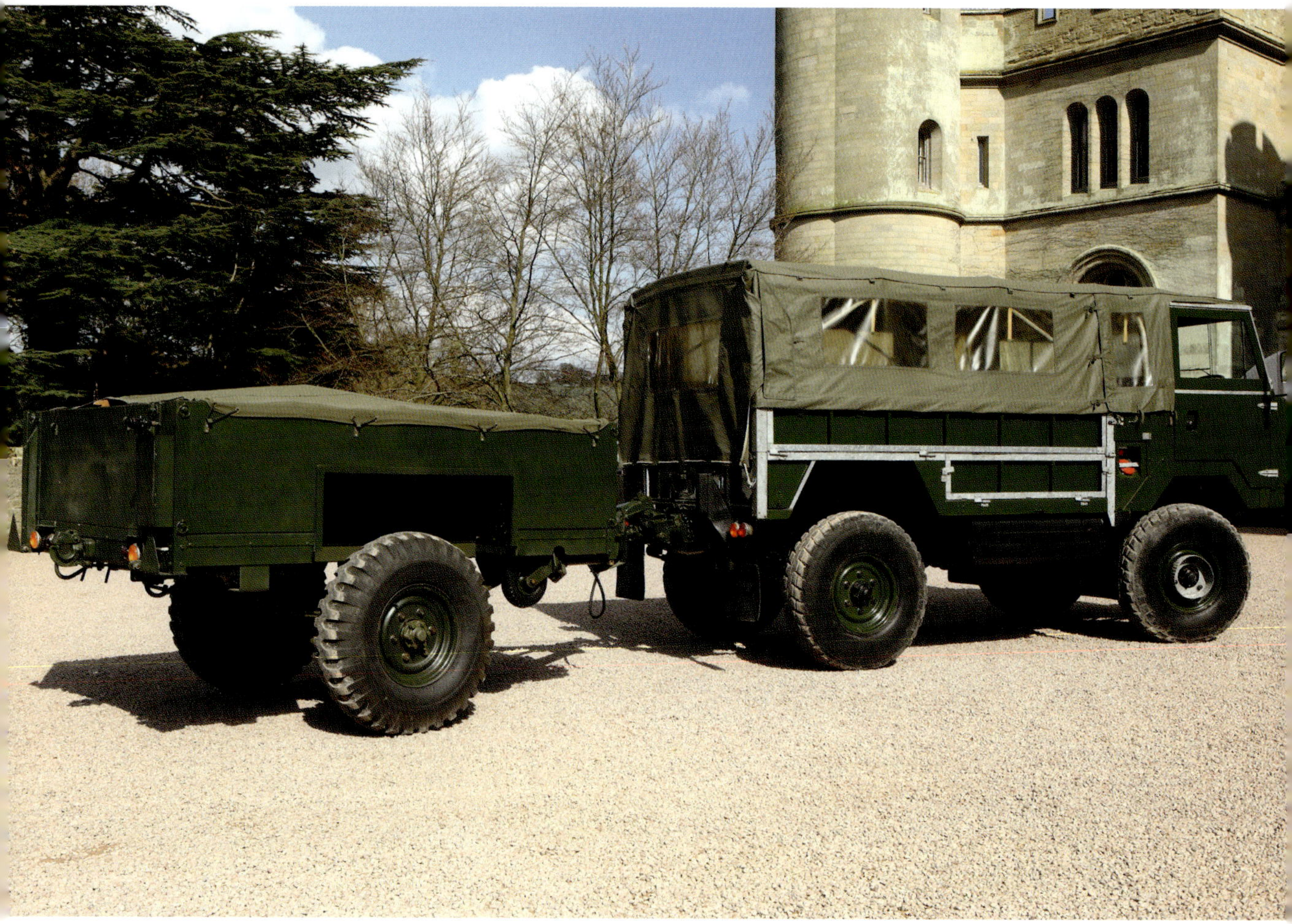

ACKNOWLEDGEMENTS

I probably started writing this book some time in the 1990s, although I didn't realize it at the time because I was simply collecting information about the One-Tonne Forward Control as it came my way. The first indication of some attempt to organize it all came in 2000, when I wrote an article for *Land Rover Monthly* magazine to celebrate twenty-five years since the 101 had entered service. That article forms the essential backbone of this book – although the story has been hugely amplified by subsequent research.

A lot of people have contributed to that research, particularly by writing to me in response to the Roverphile column that I have been running since the mid-1990s (in *Land Rover Owner* 1996–99 and 2012 to date, *Land Rover Monthly* 1999–2000 and 2010–12 and *Land Rover Enthusiast* 2000–10). Colleagues on all those magazines played their part, too, in particular Bob Morrison and Pat Ware. I couldn't compile a list of all those who have helped me with little points of detail if I tried, but those who did will know who they are.

A few people have provided special help, and I am pleased to mention them. From Land Rover itself, these have been Mike Broadhead, (the late) Norman Busby, Roger Crathorne and Bob Lees. In Australia, help has come from Glenn Smith and from several members of the Registry of ex-Military Land Rovers, notably Neil Dailey, Rod Genn and Richard Green.

In Canada, Robin Craig helped out. In Luxembourg, indirect help has come from Nico Lauryssen and Jean-Claude Thies.

In the UK, members of the 101 Forward Control Club and Register whose input proved invaluable have included Les Adams, Alan Armstrong, Peter Barratt, Darren Parsons and (the late) Chris Savidge. I must also single out for particular thanks Phil Bashall of the Dunsfold Collection of Land Rovers, military vehicle historian Geoff Fletcher, whose enthusiasm for this project matched my own, and several members of the staff at the Heritage Motor Centre over the years: Richard Bacchus, Richard Brotherton, Anders Clausager, Lisa Stevens and Jan Valentino.

To all those I have failed to acknowledge, my thanks are no less sincere. I hope they will forgive a failing memory!

© BMIHT – All publicity material and photographs originally produced for/by the British Leyland Motor Corporation, British Leyland Ltd and Rover Group including all its subsidiary companies is the copyright of the British Motor Industry Heritage Trust and is reproduced in this publication with their permission. Permission to use images does not imply the assignment of copyright, and anyone wishing to re-use this material should contact BMIHT for permission to do so.

JAMES TAYLOR
Oxfordshire, May 2014

TIMELINE

	1966	First FVRDE forward-control mock-ups
	1967	First discussions between FVRDE and Rover Company about the new vehicle
	1968	6-cylinder prototypes built
	1968 (late)	Switch to V8 engine
	1970	Comparative military trials
	1971	Contract issued to Rover Company
	1972	First pre-production models built
	1974	First four production vehicles; Trans-Sahara expedition
	1975	Start of volume production
	1978	End of production
	2000	Last 101 One-Tonne in UK military service

CHAPTER ONE

ORIGINS

The story of the Land Rover 101 began in 1965, three years before the first prototype was completed and nine years before volume production formally began, with a War Office study into its future vehicle requirements. Among other things, this study highlighted the need for a GS (general service) 4×4 vehicle with a one-ton payload that would fill the gap between the existing Land Rovers with their ¼-ton and ¾-ton ratings and the four-ton Bedford MJ and MK trucks. 'There are many roles,' the report on that study read, 'for which the smaller vehicles are inadequate and the larger one expensive or tactically unacceptable.'

In fact, the War Office already had three types of one-ton vehicle in service, but the Austins and Morrises that were rated with that capacity were elderly designs that were near the end of their service life, and the remaining Humbers had all been converted to armoured 'Pigs'. So another study was set up to develop a specification for a one-ton vehicle that would meet all the requirements that the War Office could then foresee.

One of these was that the vehicle should be capable of towing a one-ton trailer to give a total train capacity of two tons. Another was that it should be capable of towing the forthcoming 105mm light gun, which, it was already clear, would be too heavy to be towed by existing Land Rovers. At this stage, the War Office was also taking an interest in powered-axle trailers – trailers with an axle driven by the towing vehicle through its tow hitch – and the plan was to have a powered axle on the carriage of the 105mm gun. Behind a 4×4 towing vehicle, the resulting 6×6 combination was expected to give formidable cross-country ability in rough going.

At this time, there was very close liaison between the Rover Company, who built Land Rovers at their factory in Solihull, south-east of Birmingham, and

The British armed forces had been using short-wheelbase Land Rovers since the start of the 1950s, but there was a limit to what they could cram into one, as this picture of a 1960s Series IIA 88in model shows. LAND ROVER

FVRDE, the Fighting Vehicles Research and Development Establishment at Chobham in Surrey. The key figures involved were Cyril Belfitt, who was a civilian in a senior position at FVRDE, and Tom Barton, who ran Land Rover engineering. Rover was only too well aware of the importance of British military contracts to its Land Rover sales, and had established this liaison so that it could react quickly to any new military requirement that came along. It was already experimenting with powered-axle trailers, working in tandem with Scottorn Trailers Ltd, and by 1964 the Scottorn system had become commercially available as a Land Rover approved accessory.

There was more room in a 109in long-wheelbase Land Rover, which was also better suited to towing duties. One is seen here on the FVRDE test course at Long Valley, towing a field gun. There was a limit to what the 109 could tow, however, and it was not going to be capable of handling the army's new 105mm light gun. KEN TWIST, AUTHOR

Also under development in the mid-1960s was a lightweight version of the 88in model, designed to be carried under a Wessex helicopter. Some of the thinking behind this model, usually known simply as the Lightweight, would eventually carry over to the 101 One-Tonne as well. LAND ROVER

THE 110IN GUN TRACTOR

As soon as Rover learned that their existing Land Rovers were not going to be powerful enough to handle the new light gun, they got started on the design of a bigger and more powerful model that would do the job. They were clearly also aware that the new 105mm light gun was expected to have a powered carriage, and so they made sure that the new vehicle incorporated the technology needed to drive it.

As the military had not yet issued a formal requirement for a new vehicle, Rover developed their new Land Rover as a private venture. It ended up being a lot larger and heavier than existing long-wheelbase models, although it was still recognizable as a Land Rover. It was powered by a 3-litre 6-cylinder petrol engine – detuned from its Rover saloon car application – that developed 110bhp, as compared to the 77bhp of the standard 2.25-litre 4-cylinder in existing production Land Rovers. It was rather wider than the

Land Rover developed this Gun Tractor on a 110in wheelbase specially to meet the anticipated military requirement. It was designed to tow a powered-axle trailer, but it was not what the military wanted. Just two prototypes were built in 1965–66. LAND ROVER, AUTHOR

standard 109in wheelbase Land Rover and had a slightly longer wheelbase of 110in. Although Rover described it as the 'proposed British Army ¾-ton Land Rover', engineering documents show that its payload was actually a full ton.

Nevertheless, the new 110in gun tractor did not incorporate the Scottorn trailer-drive system that Land Rover had been working on. Instead, it incorporated a different system that had been drawn up by the UK military in conjunction with Rubery Owen. The full story of the Scottorn and Rubery Owen trailer-drive systems is explained in the Appendix, but it is clear that Rover had no choice at this stage but to go with the system that the military favoured if they were to win a contract.

FVRDE agreed to take a look at a prototype of this new Land Rover, of which probably only two were ever built in 1965–66. On the Rover side, there were doubts about how well the standard four-speed Land Rover gearbox would hold up behind the 3-litre engine. Earlier experiments with this engine in Land Rovers had shown the gearbox to be a weak link, and in fact the two gun tractor prototypes were later fitted with heavy duty five-speed gearboxes made by ENV.

12 ■ ORIGINS

Sadly, the stronger gearboxes did not make FVRDE any more enthusiastic about the new 110in gun tractor. One key reason was that military thinking had by this time begun to focus on a forward-control design.

THE PRELIMINARIES

This interest in a forward-control design had probably come about because the military wanted maximum carrying space in the vehicle. They had already worked out that the overall length should be around 168in (427cm), and those dimensions had reached Rover in time to guide the design of the 110in gun tractor. That length had been established so that a defined number of vehicles plus trailers could be fitted into the RAF transport aircraft of the time. The next challenge was to arrange the vehicle so that an infantry section of ten men could be fitted into it.

By the end of 1966, FVRDE had produced the first of several full-size wooden mock-ups of the vehicle it wanted. There were at least three of these and one of them was used to test whether the forward-control vehicle could indeed accommodate ten infantrymen with their full complement of equipment. Photographs show that it was a tight squeeze, but they also show mock-ups that bear a striking resemblance in general outline to the eventual production Land Rover 101. FVRDE also made a number of scale models to demonstrate the potential roles of their new forward-control vehicle, and these included personnel carrier, WOMBAT portee, four-stretcher ambulance, missile launcher and mortar carrier.

It is not clear how closely FVRDE and Rover were working together at this stage. If nothing else, Rover were clearly very quick off the mark. In May 1967

Airportability was an important part of British military thinking in the 1960s. This is a stripped-out Lightweight being carried under a Wessex helicopter of the RAF. TANK MUSEUM

This was what the British Army wanted, and this mock-up was created at FVRDE early in 1967. It bears a striking resemblance to the eventual production 101. That is a mechanical digger in the background, not a jib mounted to the vehicle! TANK MUSEUM

FVRDE followed its first mock-up with another one during 1967. Here it is being used for trials: the idea was to see if a section of ten men could fit on board with all their kit. It appears to have been a tight squeeze – and some of the kit is still on the ground behind the vehicle. TANK MUSEUM

This was another version of the **FVRDE** full-size mock-up. The spare-wheel stowage arrangement was interesting, if rather wasteful of space. TANK MUSEUM

The **FVRDE** thinking was very well-developed by the time Rover were formally asked to become involved with the project. This was a scale model of the ambulance variant they wanted. TANK MUSEUM

The British Army borrowed a Series IIB 110 Forward Control at an early stage, but trials confirmed that it was not the vehicle they wanted. Here, the trials vehicle 22 BT 25 is seen during Operation *Wagon Train*, at Lee-on-Solent, on 19 May 1967. LAND ROVER

they lent the army a production 110in Series IIB Forward Control model for trials; this vehicle was registered as 22 BT 25. At the same time, Rover returned for further assessment an experimental 112in 30cwt forward-control Land Rover that the army had recently rejected because it did not meet their latest payload requirements. Both were used during Operation *Wagon Train*, an exercise at Lee-on-Solent, which included amphibious landings. The main aim of these loans was clearly to find out how much of the existing designs could be re-used and where they fell short of the ideal.

Meanwhile, things were formalized in May and June 1967 when Rover and FVRDE got together to draw up a specification for the new one-ton forward-control gun tractor. From the beginning, it was seen very much as a joint project: even though the new vehicle was to be developed and manufactured by Land Rover, there was never any doubt that the basic design came from FVRDE, and there would be a substantial amount of military input all the way through to production. The military input was so significant, in fact, that when the Rover Company later came to offer the new Land Rover for sale to other users, the Ministry of Defence took a royalty on every vehicle sold.

A FORMAL PROJECT

So it must have been at about this time that Rover allocated the work within its engineering department. There was already a Forward Control Team in existence, and at this stage its day-to-day work was looking after development of the existing Series IIB 110in models. Leader of the team – project engineer, in Rover language – was Norman Busby. As there were no other major forward-control developments in the pipeline at this stage, it made perfect sense to give the job of developing the new Military Forward Control model to the team that had the most experience of such vehicles. Rarely appreciated, however, is that the new 101 Forward Control was simply an addition to the Forward Control Team's workload; they would continue to be responsible for updates and modifications of the Series IIB models right through until their production ended in 1972.

More than thirty years later, Norman Busby remembered how he did some initial calculations for the new model, based on a maximum length of 168.5in (428cm) set by FVRDE: 'I sat down and put four British soldiers, 22in wide, in. That was stipulated! A 10in wide spare wheel…and it left a cab a bit small for the gunners but…that's why it's a little cab!.... We want a good departure and approach angle. Draw this back down to the ground, draw the wheels in, and you get a 101in wheelbase. So we finished up with a very impressive approach and departure angle, which was helped by this short-body requirement.'

The normal project team at Rover consisted of a project engineer, an assistant project engineer and one or more technical assistants, who were usually young engineers recently out of their apprenticeship. For the new Military Forward Control, the team consisted of Norman Busby himself, his existing deputy Bob Lees as the APE and no fewer than four technical assistants: John Shaw, Scott Seymour, Kevin Hunt and Hugo Vernon. This team would be supplemented by others later on, and from time to time it would call on expert assistance from other departments, such as transmissions and electrics. The actual construction of prototypes would be overseen by Ken Twist, the foreman of the experimental shop, and would be carried out by fitters from his team.

It was standard practice in the Rover engineering department for a project engineer to issue job cards – formal requests to other engineers to carry out specific tasks – and the earliest job card for the new Military Forward Control was issued on 5 September 1967. It asked John Shaw and Ken Twist to build a cab mock-up and to assess it for ease of entry and driving position. This mock-up would have been built to the scheme drawings that must by this stage have existed, and those scheme drawings were also to be used as work started on the construction of the first prototype. A second job card, also issued on 5 September, formally requested Shaw and Twist to get started on that – although it would be some months before work started in earnest.

In the meantime, Norman Busby had wasted no time in carrying out some preliminary assessment work. The most powerful engine in the existing Series IIB Forward Control models was a 2.6-litre version of the 6-cylinder, but it was already clear that this would not be powerful enough for the new vehicle; that would have to have the 3-litre engine which had been tried in the earlier 110in gun tractor prototypes. So in July 1967, Busby had one of these engines put into a Series IIB Forward Control (335-00011A, number 110/FC/17 on the engineering development fleet). This went to FVRDE, where it acquired the registration number 00 SP 47. It is not clear whether FVRDE put it through any serious trials process, but in October or November that year it was given big 11.00 × 16 tyres and shipped out to Sharja in the Persian Gulf for trials with the Trucial Oman Scouts.

One of the things that became clear from these tests was that even the 3-litre engine would struggle to deliver the performance that the military wanted, but it was the most powerful one available to Rover at the time and the project would have to go ahead with it in the hope that a better option might become

Norman Busby was appointed to run the new Forward control project for Rover. BMIHT

available. Norman Busby had already anticipated that better option: the Rover photographic records show that a forward-control chassis was trial-fitted in June 1967 with one of the new V8 engines, which at that time were just entering production for Rover's flagship car models, the P5B 3.5-litre Saloon and Coupé.

1 TON 4×4 G.S. TRUCK (G.S.R. 3463)

By the time of the GSR document in June 1968, the Land Rover design was coming along nicely. It was reflected in this artist's impression that was included with the document. TANK MUSEUM

THE 105MM LIGHT GUN

One of the primary roles for which the 101 was intended was that of tractor for the new British 105mm light gun. Known as the L118, this was under development from 1965 by RARDE at Fort Halstead, near Sevenoaks in Kent, and was intended as a replacement for the Italian-designed 105mm Pack Howitzer that served with the British forces from 1961. Airportability was among the key requirements.

The first prototypes were tested in 1968 but production was not authorized until late 1975 and actually began at the Royal Ordnance Factory, Nottingham, in 1976. An early intention was to mount the gun on a powered carriage so that its axle would be driven from the gun tractor. However, this idea seems to have proved too complicated and by about 1973 the trailer-drive requirement had been changed so that the gun tractor would power the axle of an ammunition limber towed between tractor and gun. Within the next couple of years, this requirement was also dropped.

The 105mm light gun was originally intended to weigh no more than 1,600kg (3,520lb), but reinforcements to the gun carriage were found necessary during development and production versions weighed 1,858kg (4,88lb). The gun was 8.8m (29ft) long, and for transporting, the barrel was swung round over the trail. It had a six-person crew and could fire six rounds a minute, each shell weighing typically around 15.1kg (33¼lb) in HE (high explosive) form. The gun's maximum range was 17.2km (10½ miles) and it had an anti-tank range of 800m (2,624ft).

The British Army had fewer than 150 of these guns but they were a very great export success, being purchased by armed forces in Australia, Botswana, Brazil, Brunei, Ireland, Kenya, Malawi, Malaysia, Morocco, the Netherlands, New Zealand, Oman, Portugal, Spain, Switzerland, Thailand, the UAE, the USA and Zimbabwe. The British Army deployed five batteries of six guns each to good effect during the Falklands War, when these guns were firing up to 400 rounds a day each.

CHAPTER TWO

DESIGN AND PROTOTYPES

The design and prototype stage of the 101 One-Tonne project lasted five years, from 1967 through to 1972. It was complicated by what was really a false start, when Rover built the first group of prototypes with the old 6-cylinder engine. Once it became clear that the new V8 engine would be available for the project, a major redesign ensued and the final prototypes incorporated this engine and its associated new transmission.

A further complication was that the One-Tonne was a joint development between Rover and FVRDE. Each side had its own engineering development prototypes, although all were built by Rover. Some of these prototypes shuttled back and forth between FVRDE in Chertsey and Rover at Solihull. In addition, as Rover had hopes of selling the 101 to overseas military agencies once the MoD production specification had been signed off, there was a demand for demonstrator vehicles from quite an early stage. On more than one occasion, the Rover development team found itself short of vehicles.

INVITING TENDERS – 1968

The specification that Rover and FVRDE had worked up between them was turned into a formal requirement document, GSR 3463, in June 1968. (Those initials stood for General Staff Requirement, and it was standard procedure for the War Office to issue such a document when it was looking for a new item of equipment.) This document described a forward-control 4×4 vehicle, and the accompanying illustration looked astonishingly like the eventual production Land Rover 101, even though the contract for production vehicles had not yet been awarded to Rover.

The vehicle that the War Office wanted had a payload of one tonne, metric measurements by this stage being the norm in the British military. As the metric tonne is lighter than the Imperial ton (2,204.5lb as against 2,240lb), this actually reduced demand on the vehicle. Its overall length was the 168.5in (428cm), which would allow it to fit into Andover transport aircraft, and the military wanted an overall weight of 7,700lb (3,493kg), achieved if necessary by removing non-essential elements of the vehicle. This weight limit would allow it to be airlifted by the latest Wessex heavy lift helicopters. It had to have the powered-axle trailer capability, with a detachable propshaft driven via the rear tow hitch; and the trailer capacity had now increased to 1.5 tonnes, no doubt in order to cope with weight of the 105mm gun.

That GSR document, now known by its military code of LV82, was issued to the motor industry at large in August 1968 – by which time Rover had been working on the vehicle it described for the best part of a year! The document recognized there were no fewer than sixteen vehicles in the frame for the eventual military contract, but when bids were submitted in September 1968, only two manufacturers confirmed their readiness to tender for the contract. The two were Land Rover, with the 101, and Volvo, with a version of the 4140-series Forward-Control that they were already supplying to the Swedish Army in quantity. Proud of their bid for the contract, Land Rover displayed a 101 prototype at the Commercial Motor Show at London's Earls Court during the autumn.

THE 6-CYLINDER PROTOTYPES – 1968

The chassis design of the existing forward-control models was something of a compromise because it was a modification of the chassis drawn up for the 109in normal-control models. Essentially, a forward extension had been added to support the cab, and a sub-frame had been added at the rear to support the body. With the new 101in Forward Control, the chassis was designed from first principles, and it ended up as a simpler, straight-frame design.

The first six 101 prototypes were built with the 3-litre 6-cylinder engine, plus the four-speed primary gearbox and special low-ratio two-speed transfer gearbox used in the 110in Series IIB Forward Control. This combination gave selectable two-wheel drive, and to cope with the transmission loadings the 101 prototypes also had heavy duty ENV axles from the Series IIB Forward Control.

All of the 6-cylinder prototypes were built with more or less the same configuration, although the specification developed in detail as Rover's own test work proceeded. They were all boxy looking machines with flat panels and a rather crude-looking snub nose, and some had separate canvas tilts for the cab and the back body, while others had a one-piece tilt. The upper bodywork was made demountable to reduce the height so that the stripped-down vehicle could be fitted into an Andover transport aircraft. This was something proposed by FVRDE that had

ABOVE: **The first six prototypes all had the Rover 3-litre 6-cylinder engine. Here it is in the first prototype, 101/FC/1, as that was being built at the start of 1968.** BMIHT

RIGHT: **Not only was the length influenced by the space inside an RAF Andover transport aircraft. This picture, from early February 1968, shows 101/FC/1 stripped down for airporting operations, with most of its upper superstructure removed. The wooden arch represented the Andover's hull.** LAND ROVER

Prototype 101/FC/4 was visually typical of the 6-cylinders. It was pictured here in summer 1972, complete with powered-axle trailer, while still undergoing assessment with the army. LAND ROVER

been demonstrated on one of their early scale models. The principle was similar to that being employed for the 88in Lightweight Land Rover that was under development at the same time, although for that vehicle the upper bodywork was made demountable to reduce weight for helicopter lifting operations.

The joint nature of the 101 project became very clear when these prototypes were built. Rover retained just one of the six for engineering development, and constructed one specifically as an FVRDE prototype on which the military could carry out their own development. This and the remaining four fulfilled FVRDE contract WV7791 for five prototypes ('engineering models' was the term used) that was issued to Rover on 16 January 1968.

The first vehicle completed became the Rover engineering prototype and was numbered 101/FC/1; it was built between February and mid-May

This was 101/FC/1 as built. Although the essential structure was there, the 101 would undergo a huge number of changes before production began. TANK MUSEUM

1968. The last vehicle, 101/FC/6, became the FVRDE engineering prototype and was built between August and December that year. Certain items of its specification, perhaps including a lightweight chassis, were special. The other four prototypes, 101/FC/2 to 101/FC/5, were destined for FVRDE trials, and were built between March and October. They were registered with military numbers 01 SP 13 to 01 SP 17, in the SP series used for experimental vehicles.

Three of these first six vehicles (numbers 1, 5 and 6) had a 24V electrical system, and three had a 12V system (numbers 2, 3 and 4). All except numbers 2 and 3 were fitted with Mayflower winches, mounted at the front of the vehicle between the chassis rails and in front of the radiator. All except number 6 had twin fuel tanks, one on either side of the chassis. Three of the FVRDE group were sent on to the Royal School of Artillery (numbers 3, 4 and 5), mainly for trials associated with the forthcoming 105mm gun, and two (numbers 2 and 6) remained at FVRDE for other trials.

Four of the 6-cylinder prototypes had a front-mounted Mayflower drum winch. The date of this photograph – early February 1968 – makes clear that this must be the first prototype under construction. LAND ROVER

AUSTRALIAN INTEREST

Meanwhile, Land Rover had been busily making clear to some of their military customers that the 101 was under development and would be ready for sale in due course. Among those who took the bait were the Australians, who had a requirement for a new one-ton GS vehicle. They really wanted a larger vehicle than the 101 because they used a huge stores

THE 3-LITRE 6-CYLINDER ENGINE

The 3-litre 6-cylinder engine used in the first 101 prototypes was a detuned version of the Rover car engine that had entered production in 1958. With a swept volume of 2995cc (77.8mm bore by 105mm stroke), an 8:1 compression ratio and a single SU carburettor, it developed 110bhp at 4,500rpm and 150lb ft of torque at 2,000rpm. The engine was a direct descendant of the 2.1-litre Rover car engine introduced in 1948, itself a 6-cylinder version of the legendary 1.6-litre engine that had powered the first Land Rovers.

From 1959, the 3-litre engine began to appear in experimental Land Rovers that were going to be big and heavy or were expected to have a high payload. However, all these vehicles confirmed that the existing Land Rover gearbox was not strong enough to take the engine's torque, and this was the main reason why the short-stroke 2.6-litre derivative was chosen to power the forward controls from 1963 and the normal-control models from 1966.

Nevertheless, the 3-litre was considered for various other applications where a special gearbox could be fitted to cope with its high torque output, and heavy duty ENV gearboxes were tried in the 110in gun tractor, in the 112in and 120in forward controls, and in the specially modified 110in forward control that was trialled by the Trucial Oman Scouts. Perhaps the 101 team anticipated switching to such a gearbox at a later stage, although it is rather surprising that they did not try one out in any of their 6-cylinder prototypes.

DESIGN AND PROTOTYPES   ■ 21

Prototype number 7 was built as a demonstrator for the Australian military. The Australians were interested in using a locally made Ford engine, which is seen here during the construction of 101/FC/7 in October 1968. BMIHT

THE FVRDE 'BROCHURE'

It must have been in late 1968 or early 1969 that FVRDE produced its own 24-page 'brochure' for what it called the Truck, 1 Ton, 4×4 General Service Forward Control. This was probably intended for distribution to military personnel who might have an interest in the latest project, and resumed the position.

No mention was made of Land Rover's involvement, although there was no attempt to disguise the Land Rover grille badge on the 'engineering model' that was pictured. This was in fact 101/FC/1. The brochure focused on the vehicle's airportability and mobility. It noted that the new forward control had a 65 per cent larger platform space than the then-current Land Rover 109, and that its 1,016kg (2,235lb) payload capacity was likely to be increased to 1,270kg (1.25 Imperial tons or 2,794lb). It also noted that a hardtop could be fitted for command post, workshop or radio station use, and that this brought with it a weight penalty of 300lb (136kg). FFR versions were envisaged, together with a powered-trailer drive, and both Terra-Tires and sand tyres could be fitted with minimal modification.

and equipment pallet, but Land Rover persuaded them to take a look at the 101, as it was then. However, the Australians did insist that whatever vehicle they bought should have a locally made engine. The one selected as suitable for the 101 was a 221cu in (3621cc) Ford Falcon 6-cylinder engine with 135bhp.

Bob Lees, the assistant project engineer, took charge of this special project. He remembers that Land Rover put forward three separate proposals. The first was that the Australians should take a Falcon-engined, but otherwise standard, 101. This was known to the Australians as the DDR type. The second was that they should take a widened 101, still with the Falcon engine. This was known as the DDX type. The third was that Land Rover should build a vehicle to a design that the Australian military vehicle research establishment had drawn up.

So a seventh 101 prototype was built to meet that first requirement at the end of 1968, and its Ford Falcon engine was installed during the build at Solihull. Prototype 101/FC/4 was also shipped out to Australia for comparison trials during April and May 1969. The grand plan was that Lees would go to Australia to supervise local assembly of the Australian 101s if Land Rover had won the contract, but they did not. (There is more information about the Australian trials in Chapter 6.)

THE TURNING-POINT

The last of the 6-cylinder prototypes were still under construction in late 1968 when the planned specification of the 101 went through some major changes. Over the next year or so there would be no fewer than three of these and all would have a fundamental impact on the eventual production specification.

The first of these changes was the replacement of the 6-cylinder engine by Rover's new V8 engine. That was accompanied by the arrival of the new gearbox designed to go with it, and this second change brought permanent four-wheel drive in place of the original selectable four-wheel drive. The third change, which arose out of the change to a permanent four-wheel drive, was to new axles with six-stud wheel fixings – the only production Land Rover axles to have six rather than five wheel-fixing studs.

The 101 team must have been casting envious eyes in the direction of the V8 engine for some time, not least because military trials with the 6-cylinders had resulted in an alarming number of engine failures. The new 3.5-litre V8 had entered production at Rover during 1967, but all the initial production capacity had been earmarked for Rover saloons, and there was no point in building prototypes with an engine that would not be available for production. Fortunately, by late 1968 or early 1969, V8 production had built up to such an extent that it did look as if there would be some spare capacity that could be devoted to the 101.

The V8 engine had also been chosen as the power unit for the new Land Rover 100in Station Wagon that was then under development, and that would enter production in 1970 as the Range Rover. For that vehicle, a special configuration with a raised water pump had been developed, and this would suit the 101 very well. Again, for the Range Rover, Frank Shaw's transmissions department had begun work on a new heavy duty four-speed gearbox with integral two-speed transfer box, and the 101 team was able to piggy-back on this development.

Many years later, in 1995, Frank Shaw remembered that there had to be compromises in the development of that gearbox:

The problem was, we had two vehicles but we could only have one gearbox. We'd got the money to build one gearbox. [They] persuaded me to make the gearbox suitable for the forward-control vehicle, and it was really too big for the Range Rover.... One of the things I did on the Range Rover gearbox to suit [the 101] was to make a transfer that would give you the same speed of output on the power output and the axle, for towing a powered trailered gun. And that caused a little bit of complication that wouldn't normally have been there.

In practice, the 101 and Range Rover gearboxes would be built as two variants of a core design. The 101 version would incorporate two power take-offs which were not present on the Range Rover version;

With the switch to the V8 engine came the new gearbox and permanent four-wheel drive system that had been designed for the Range Rover. This picture shows the old Land Rover gearbox (later known as an LT77) and transfer box with selectable four-wheel drive on the left. The new LT95 gearbox with integral transfer box is on the right. The LT77 has a transmission brake installed. LAND ROVER

one was for the powered-trailer drive, and the other (a 'bottom' PTO) would be used to power the winch.

The new gearbox was necessary in order to handle the torque of the V8 engine, but with it came a new departure. Putting all the V8's torque through the rear axle in a selectable two-wheel-drive system demanded a weighty, heavy duty axle that spoiled the ride quality. However, Range Rover Lead Engineer Geof Miller had suggested using a transfer box that gave permanent four-wheel drive. This split the torque between two axles, which meant that both axles could be lighter-duty types and consequently lighter in weight, and that made a major improvement to the ride quality. So the new gearbox, known as the LT95 type, was drawn up to give permanent four-wheel drive through its integral transfer box. The need for better ride quality in the Range Rover thus indirectly influenced the whole future of the 101 project.

The third major change was to the 101's axles. The first seven prototypes – and the eighth, which would be the first one with a V8 engine – all had heavy duty ENV axles with Land Rover's standard five-stud wheel fixings. Saving weight was always important to the 101 project, as the vehicle was destined to be airportable, and so the 101 development engineers seized on the opportunity to use lighter axles. The ones they chose were made by Salisbury, and with these came a six-stud wheel fixing, which the Land Rover engineers saw no need to change.

THE FIRST V8 PROTOTYPES – 1969

The MoD obviously had no objection to getting a vehicle with a more powerful engine. Even in its 8.5:1 low-compression form, the V8 engine offered more power and vastly more torque than the old 6-cylinder. The actual figures were 135bhp as against 110bhp, and 205lb ft of torque as against 150lb ft. Rover's next job was clearly to build a V8-engined prototype.

The story of that first V8-engined 101, the eighth prototype, is immensely complicated. When Norman Busby issued a job card requesting its construction on 16 July 1968, it was to be a left-hand drive 6-cylinder type, with 24V electrical system, Mayflower winch and powered-trailer drive. It was also to be

> ### ROVER AND THE V8 ENGINE
>
> The 3.5-litre V8 engine was not a Rover design, but had originated with the Buick division of General Motors in America. A series of incidents led to Rover's acquisition of the manufacturing and development rights to this engine, which had been hailed as a revolutionary design on its introduction in 1960 Buicks, Oldsmobiles and Pontiacs. The key factor was Rover's difficulties in selling Land Rovers in the USA, and the company's North American representative Bruce McWilliams was sure part of the difficulty was that the vehicles were underpowered for local expectations. He secured agreement from Rover's Managing Director, William Martin-Hurst, to find a suitable American V8 for use in American-market Land Rovers.
>
> McWilliams already had his eye on one candidate when Martin-Hurst himself stumbled across the GM V8. Ideal in dimensions (it would fit into existing Rover cars as well as Land Rovers) and weight (block and cylinder heads were made of aluminium alloy), it was also a very modern design that was capable of further development. General Motors had taken it out of production in 1964 when they had found a way of making lightweight iron-block engines more cheaply, so they were happy to find a buyer.
>
> Martin-Hurst opened negotiations with GM as soon as he could and secured the engine for Rover by January 1965. Rover redeveloped it to suit UK manufacturing methods and UK ancillaries, and it entered production for Rover saloons in mid-1967. Ironically, in view of the original reasons for its acquisition, Rover could never build the engine in sufficient quantities to make it available in Land Rovers as well, before 1979, although in the 1970s it was used in the Range Rover and, of course, the 101 Forward Control.

used for a demonstration or trial to the South African military (which is odd, because South Africa is a right-hand drive country), but this part of the plan was later cancelled. Nevertheless, a pre-production 101 did go out to Leykor in South Africa some four years

LEFT: The first prototype built with a V8 engine was 101/FC/8, and this also had the new front-end design. Nevertheless, there were still many differences from the eventual production specification. This view of 101/FC/8 in the workshop at Solihull shows that it was built with the fuel filler mid-way along the body side, while the fire extinguisher was mounted diagonally on the left-hand rear panel. A wheel step was now envisaged to aid access to the cab, and the production style of canvas was in place, with a curtain over the spare wheel that was behind the driver. LAND ROVER

BOTTOM LEFT: This may well be 101/FC/8 again: note the mark on the left where something (possibly the fire-extinguisher mounting) has been removed. The cushions of the inward-facing seats have stepped edges, which did not go into production, and the ducting visible above the engine cover appears to be an early attempt to get heat into the back body. The spare-wheel mounting arrangement is still not the production type. LAND ROVER

later, and this was presumably the fulfilment of the original South African interest in the new model.

Construction of 101/FC/8 probably began in November 1968, and the vehicle was fitted with a V8 engine; it had probably switched from left-hand drive to right-hand drive by this stage. By January 1969 it was substantially complete and was being used for cooling tests. Then in March 1969, Norman Busby gave instructions for it to be re-worked and completed. By July 1969, the vehicle was to be given a lightened chassis frame and a new cab, and by this stage it was so different from the vehicle originally planned that it took on the new designation of 101/FC/8A. It appears to have been finally considered finished early in September 1969.

So 101/FC/8A was not only the first 101 to be built with the V8 engine, LT95 gearbox and permanent four-wheel drive, but it also had five-stud ENV axles like the 6-cylinder models, presumably as the lighter six-stud axles were not yet available.

Photographs show it with the early style of two-piece canvas tilt. It also had another new weight-saving development – taper-leaf springs, which were more expensive but also more durable than the semi-elliptic type. Norman Busby remembered that these saved around 17lb (7.7kg) weight at each corner of the vehicle. Those used on 101/FC/8 were shorter than the ones later adopted. This eighth prototype was also the first vehicle to be built with a different type of winch. Instead of the front-mounted Mayflower winch driven from the nose of the crankshaft, it had a Nokken winch mounted on the left-hand chassis side member.

This new winch had several advantages over the Mayflower type. First, it was considerably lighter in weight – and weight was always a primary military consideration on vehicles destined to be airlifted, as the 101 was. Second, in its position underneath the body at the side of the vehicle, it could be driven from a power take-off mounted on the gearbox. Third, its cable could be fed to either the front or the rear of the vehicle, so giving double the versatility of the Mayflower installation. Fourth, deletion of the front-mounted winch meant that the chassis could be shortened at the front end, so reducing the vehicle's approach angle. One subsidiary result was that the front end of the bodywork could be redesigned, and a new and less brutal design with sloping, instead of square, panels was used. The front bumper design also changed – although its final design was still some way in the future.

There was a problem, however. Always short of prototypes – of the seven 6-cylinders built, only one had been retained at Solihull for development – Norman Busby could see that a single vehicle to the new V8 specification was not going to be enough. Unfortunately, with eight vehicles now constructed, he was being told by his management that he had enough and that there was no budget available to allow him to build any more.

So Norman did the best he could with what he had. He arranged for 101/FC/1 to be converted to the latest V8 specification. The job card for the re-work was issued on 21 November 1968, and it asked for a chassis frame similar to that of 101/FC/8 (the V8 prototype then under construction), with a single 24-gallon (109ltr) fuel tank on the left-hand side and lightweight cross-members and outriggers like those on 101/FC/6. Work was to begin as soon as 101/FC/8 was completed, and there was certainly a degree of urgency about it. FVRDE were expected to 'clarify' the one-tonne specification in February 1969, and 101/FC/1 was to meet what Rover thought this revised specification would require. In practice, 101/FC/1 seems to have been rebuilt during February 1969, so being completed as a V8 even before 101/FC/8A.

This is the first prototype, 101/FC/1, in the Rover experimental department in summer 1969 after conversion to V8 power. The front end displays a 'transitional' design. Note the two large windows in the side of the hood, a feature probably unique to this vehicle. LAND ROVER

This was 101/FC/8 as first built, with a 'transitional' front-end design. LAND ROVER

Even then, Norman Busby knew that he would need more vehicles for the project, and he had to do what he described many years later as 'a bit of wheeling and dealing' to get the extra vehicles he needed. As he explained in 1996, 'People were always leaning on me: "Why have you got so many vehicles?"' In the end, he managed to get two more prototypes: one by constructing a vehicle from spare parts that had already been accounted for in the project's budget and the other by getting special permission from Rover's managing director, A. B. Smith, during 1970.

Both of these prototypes were built during 1971, the 'permitted' vehicle becoming 101/FC/9 and the 'spare parts' vehicle becoming 101/FC/8C. That '8C' designation was all part of the concealment ploy: a casual look would suggest that this was the eighth prototype, when in fact it was the ninth. As for the C in its designation, when 101/FC/8 had been rebuilt for the second time in spring 1970, it had been re-numbered 101/FC/8B. Even today, working out the story of the prototypes with an 8 in their numbers is not easy!

Both of these final prototypes also incorporated a number of differences from the earlier V8s, the most obvious being a revised cab design that had been pioneered on 101/FC/8B, and a one-piece canvas tilt. To a casual observer, they looked like the eventual production vehicles of three years later – but there would be a whole lot of detail changes before production began at the end of 1974.

THE TRIALS – 1970

Back in late 1969, however, Rover's primary focus was on providing 101s to the latest specification for the formal military trials that were scheduled to begin during 1970. The whole process was highly unusual because FVRDE had already been running five 101 prototypes for more than a year by the time of the trials, but they were, of course, 6-cylinder types to the earlier specification. This time the objective was to compare the latest V8-powered vehicle head-to-head with its only real rival for the contract, the Volvo 4141 Laplander, in a standard competitive reliability trial.

Although it appears that FVRDE had already asked Rover for a total of six vehicles, all with powered-axle trailers, FVRDE – newly renamed as MVEE – was satisfied with just two for the trials. That was fortunate: Rover had only 101/FC/1 and 101/FC/8 (by now 101/FC/8B) equipped with V8 engines, and so it was these two that upheld the honour of the 101 against the Volvo. The other four vehicles were wanted for user trials, which would be conducted separately at a later date.

The trials were conducted at the Longcross military test track near Chobham, and Norman Busby remembered that 'we had a very large tent…our fitters virtually lived in it at Chobham. We had a team of people going down and coming back again with bits and pieces, and staying for a week and coming

ABOVE: **101/FC/8B, now in its final form, was pictured at the Longcross test track with its rival for the MoD contract, the Volvo 4140.**

RIGHT: **This is the V8 engine in 24V FFR guise, pictured over the summer of 1970. The big 90A alternator is clear in this view. The host vehicle may be 101/FC/8B – one of the oddly-numbered prototypes used to get around budget restrictions.** LAND ROVER

back. Ken Twist [Foreman of the Experimental Workshop] was a key factor in those trials'. Both the 101 and the Volvo were displayed during the course of the trials in mid-1971 at the annual military vehicles exhibition held by the MVEE at Chertsey in association with the Society of Motor Manufacturers and Traders. They were described in the exhibition catalogue as 'under assessment'.

Even though the specification of the 101 had been drawn up in such close liaison with its planned users, the Land Rover did not score a walkover victory. The Volvo was a well-tried vehicle that proved to be a strong competitor, and it had an amphibious capability that the 101 did not. The manufacturers were permitted to make changes during the trials as problems showed up, and Norman Busby remembered that 'there were one or two minor adjustments on chassis outriggers, cross-members' during this period. But eventually the 101 satisfied MVEE's testers and Land Rover was awarded the contract, number WV9615, on 3 December 1971. Rover – under the aegis of British Leyland – announced the fact in public at the Commercial Motor Show in autumn 1972.

MALAYAN INTEREST AND THE TERRA-TIRES

However, the story of the 101 prototypes was not over yet. One reason why Norman Busby had wanted those two extra prototypes at the end of 1970 was that there was further interest from overseas in the new 101 Land Rover. If demonstrators were required, he could see that he might be left with no vehicles to work on at Solihull – as had almost happened during the 6-cylinder stage of the project.

The Malayan military already had a large fleet of Series IIB Forward Control Land Rovers and understandably wanted to know more about the forthcoming 101. So Rover borrowed 101/FC/4 back from FVRDE – it was now redundant because the specification had moved on – and shipped it out to Kuala Lumpur as a demonstrator during 1970.

Meanwhile, the MoD had already shown an interest in Goodyear Terra-Tires, which were huge swamp tyres that provided maximum flotation over boggy ground. Rover had fitted a set to 101/FC/6 in March 1969, and articulation tests – presumably also with 101/FC/6 – were carried out in June. Then in April 1970 the MoD took a more active interest, borrowing the two 110in gun tractor prototypes and testing them with both Terra-Tires and powered-axle trailers.

The Malayans then expressed an interest in trying Terra-Tires, so in 1971 Rover sent 101/FC/1 out for more trials in Kuala Lumpur with Terra-Tires and a powered-axle trailer. Rover agents Champion Motors acted as intermediaries for the company. Endless

101/FC/1 went to Malaya for trials in 1971, and was equipped with Terra-Tires. The trailer was the one that had been used with 101/FC/8, and still carried that vehicle's number plate.
LAND ROVER

FVRDE AND MVEE

During the development of the 101 Forward Control, there were organizational changes on the military side.

FVRDE, the Fighting Vehicles Research and Development Establishment, had been created in 1952 from two earlier military organizations: one involved with vehicle design and the other with vehicle testing. It was based at Chertsey in Surrey, and used test facilities at Longcross (near Bagshot) and Long Valley (Aldershot).

In April 1970, FVRDE amalgamated with MEXE (the Military Experimental Engineering Establishment) that had been based at Christchurch in Dorset. The organization that emerged was known as MVEE, the Military Vehicles and Engineering Establishment, and had its headquarters at Chertsey.

The odd front bumper confirms that this is indeed 101/FC/1 after rebuild to V8 power, as the number painted on the spare wheel rim suggests. An alternative mounting position for the spare wheel was clearly under consideration. LAND ROVER

confusion has resulted from photographs showing the powered-axle trailer wearing the registration plate of HXC 805H that belonged to 101/FC/8, but 101/FC/8 was not sent to Malaya. It was simply that the trailer formerly used with it was sent, and nobody bothered to remove the number-plate it was carrying at the time.

The trials seem to have gone well, and Norman Busby remembered that the Malayans were keenly interested to buy some 101s. However, as the British military trials had not yet been completed, there was no final production specification, so Solihull's sales people asked the Malayans to wait before placing an order. Various delays in the production of the 101 over the next few years meant that the Malayans ran out of patience and decided to go elsewhere for their new vehicles. The Terra-Tire option was not pursued further for the 101.

THE 101 PROTOTYPES

There were 10 prototypes of the 101 One-Tonne.

101/FC/1 (Unregistered; FVRDE Wing Number 7888L)

The first prototype was built between February and mid-May 1968 as a 6-cylinder vehicle with 24V electrical system, a front-mounted Mayflower winch and a powered-trailer drive. It had separate canvas tilts for cab and back body.

This vehicle belonged to the engineering department at Solihull, and appears never to have been road-registered. It must have run on trade plates when it was away from the factory. After sale into private ownership, it was registered as JKV 407N.

101/FC/1 was still on the engineering fleet in early 1969, and was the only example of the 101 then at Solihull; the next five prototypes had all gone to military units for trials, and the seventh prototype had gone to Australia. It was converted to V8 power in early February 1969 and was rebuilt with a 'transitional' style of front end that anticipated the production design. It lost its Mayflower winch at that point. Later that year, it was put through a 5,000-mile endurance test at FVRDE. In early 1970, it went to FVRDE with 101/FC/8 for comparative trials against the Volvo 4140, and while there was mocked-up with a second steering wheel and instrument panel on the left-hand side.

In August 1970, 101/FC/1 was fitted with Goodyear Terra-Tires and went to Malaya for evaluation trials. On that occasion, it was used with a powered-axle trailer that had earlier been used with 101/FC/8 and still carried that vehicle's registration plate reading HXC 805H.

101/FC/2 (01 SP 13; FVRDE Wing Number 7666)

The second prototype was built as one of the four trials vehicles for FVRDE in March–June 1968. It was a RHD 6-cylinder with a one-piece canvas tilt, 12V electrical system and trailer drive, but no winch. The engine number was 3L-LC8-7 (3-litre, low compression 8:1, number 7.)

The vehicle was delivered to FVRDE and carried Wing Number 7666. In August 1969, it was fitted with a wooden mock-up of a proposed field ambulance body. The vehicle was struck off on 3 June 1975 and was sold that July, but has now been lost.

101/FC/3 (01 SP 14; FVRDE Wing Number 7688)

The third prototype was another trials vehicle for FVRDE and was built between May and August 1968. It was another 12V RHD vehicle with 6-cylinder engine and trailer drive. Although it had a winch-type bumper with cutout for the fairlead, no winch was actually fitted. There was a one-piece canvas tilt. The engine number was 3L-LC8-8.

101/FC/3 was delivered to FVRDE, who sent it on to the Royal School of Artillery at Larkhill. In April 1970, it became the first 101 to be fitted with a winterization kit. During 1972, it did a 10,000-mile trial at MVEE. It was probably after this that 101/FC/3 was equipped with the palletized version of the Swingfire anti-tank missile that was then under development. (This later became known as Beeswing and was marketed as Infantry Swingfire.)

Like 101/FC/2, this vehicle was struck off at Chertsey on 3 June 1975. When sold into civilian hands, it went to Jersey, where it worked on the beach for a local council. On return to the mainland in 1988, it acquired registration number OAB 266P. The vehicle survives in private ownership, and has been meticulously rebuilt.

101/FC/4 (01 SP 15; FVRDE Wing Number 7698)

The fourth prototype was constructed between May and September 1968. It was a 12V vehicle with RHD, the 6-cylinder engine, trailer drive and a Mayflower winch. It had separate tilts for cab and back body, like 101/FC/1. The engine number was 3L-LC8-12.

This was the third trials vehicle for FVRDE, but it appears to have spent little time there. It joined 101/FC/3 and 101/FC/5 at the Royal School of Artillery, and was subjected while there to a 2,000-mile cross-country test at GVW.

The Rover Company borrowed it back twice. On the first occasion, it accompanied 101/FC/7 to Australia for gun trials between April and May 1969. On the second occasion, in August 1970, it went to Malaya as a demonstrator. It appears that this demonstration visit was separate from the Malayan trials that involved 101/FC/1.

Its eventual fate is not known.

101/FC/5 (01 SP 16; FVRDE Wing Number 7708)

The fifth 101 prototype was built between June and October 1968 as a RHD 6-cylinder vehicle with 24V

101/FC/3 was built with a 12V electrical system and a winch, and went to the Royal School of Artillery for assessment. It was pictured here undergoing trials as a Beeswing missile launcher; one of the missiles is missing from the array. Just visible is the tubular framework of the cab hood, here folded up. TANK MUSEUM

DESIGN AND PROTOTYPES 31

electrical system, trailer drive and Mayflower winch. Like 101/FC/1 and 101/FC/4, it had separate tilts for cab and back body. The engine number was 3L-LC8-10.

It was delivered to FVRDE and then joined 101/FC/3 and 101/FC/4 at the Royal School of Artillery. It was displayed at Artillery Day on 18 July 1970. At Larkhill, it was put through a 5,000-mile cross-country test and a 5,000-mile road test at GVW. It was later used for further Beeswing missile system trials.

101/FC/5 was struck off on 20 July 1973 and sold that November. When it passed into civilian hands, it gained the registration number XRC 779M. By 1990, it was owned by Ampleforth School in Yorkshire and is believed to have had a V8 engine. Whether that engine had been put in at Rover before sale or was a later conversion is not clear.

101/FC/6 (01 SP 17; FVRDE Wing Number 7814)

This was the last of the original 6-cylinder prototypes, and was built as a development vehicle for FVRDE between August and December 1968. It was built as a standard GS vehicle with 6-cylinder engine (number 3L-LC8-11), 24V electrical system, trailer

This is almost certainly 101/FC/6, pictured here at Towcester in April 1969. Although the vehicle carried no registration number at the event, its powered-axle trailer did – VXC 100F was the number of one of the 110in Gun Tractor prototypes that had been tried with the trailer!
LAND ROVER

101/FC/6 was later modified with a box body, and survives in that form. The textured finish to the front bumper was unique to the vehicle.
AUTHOR

drive and Mayflower winch. It had separate tilts for cab and back body. Unlike the other prototypes, it had just one fuel tank (on the left of the chassis) rather than twin tanks. It may well have been built with lightweight cross-members and outriggers, which in their initial form led to chassis cracking on test. It also had a unique front bumper finish, with a hatched pattern in the metal. The purpose of this is unclear.

101/FC/6 went to FVRDE, but probably spent most of its life shuttling between Chertsey and Solihull. In March 1969, a vehicle wearing the registration plate 01 SP 17 was photographed at Solihull wearing Terra-Tires, but this may not have been 101/FC/6. In April 1969, it was probably 101/FC/6 that was damaged during a demonstration at Land Rover's twenty-first birthday event at Towcester, and was subsequently rebuilt.

Some time in 1969 or 1970, 101/FC/6 went to Marshall's of Cambridge to be fitted with a box-body. The plan was to explore the possibilities of using the 101 for specialist signals and electronics duties, in particular to replace the Austin one-ton trucks that then fulfilled those roles.

After its life as a trials vehicle was over, 101/FC/6 vehicle went to the vehicle collection at the Beverley Museum of Army Transport. It now belongs to the REME vehicle collection at Bordon.

101/FC/7 (Registration Not Known)

The seventh prototype was specially built with a 3.6-litre 6-cylinder Australian Ford Falcon engine for Australian military trials. It was constructed between September and December 1968 and was equipped with a powered-trailer drive system and a Mayflower winch. It had a 24V electrical system with 60A alternator, separate tilts for cab and back body, a 9.5in clutch, and twin 18-gallon (82ltr) fuel tanks. The axles were ENV types with a special low ratio.

It went to Australia, in April 1969, accompanied by 101/FC/4. The trials occupied both April and May that year, but were not followed by an order. 101/FC/7 remained in Australia and was subsequently sold to a mining company in Tasmania. It was later fitted with a Chevrolet V8 engine and an 'outback tour' bus body. It still survived at the time of writing, in this much-modified condition and in private ownership, in Australia.

101/FC/8 (also 8A and 8B) (HXC 805H; FVRDE Wing Number 7920L)

The story of the eighth prototype is a very complicated one. It was planned as a LHD 24V vehicle with the 6-cylinder engine in 1968, but between December 1968 and January 1969 became the first 101 to have a V8 engine. Still incomplete at this stage, it was rebuilt in mid-1969 with a lightened chassis frame and modified cab and front end to reflect the latest design thinking. As completed, 101/FC/8 had a unique specification, with front-mounted winch, V8 engine and powered-trailer drive system. Like 101/FC/6, it was built with just one fuel tank, on the left of the chassis.

101/FC/8 was road-registered on 10 September 1969 as HXC 805H and was re-designated as 101/FC/8A, although some job cards still called it 101/FC/8 and the entry in the Solihull registration records also calls it 'FC8'. Between September and November 1969, it went through a 5,000-mile endurance trial at FVRDE Bagshot.

It was rebuilt yet again in January 1970, to the proposed FVRDE specification, with the spare wheel now located behind the cab and the cab otherwise modified. At this stage, its single fuel tank was relocated on the right of the chassis to make room for a Nokken winch on the left. The fuel filler, uniquely, was located just ahead of the rear wheelarch, and the vehicle still retained its original five-stud axles. In early 1970, it went with 101/FC/1 to FVRDE for trials against the Volvo 4140. On return to Solihull, it became a static vehicle and was used for installation tests for a few months.

It was made mobile again in April or May 1970 and was further rebuilt, this time attracting the engineering fleet number 101/FC/8B. It retained its registration number of HXC 805H and was now fitted with six-stud axles.

By 1971 it had its fuel filler in the production position behind the cab, and it also had a fixed cab window with a small opening section containing sliding Perspex. In this guise, it was shown at the 1971 SMMT–British Army Equipment Exhibition. By December 1971, it had become a static installation and test vehicle again, but was rebuilt as a runner soon afterwards for loan to MVEE. It was then used for Rapier installation trials and, over a period up to about mid-1974, it covered a high mileage, towing a

DESIGN AND PROTOTYPES

This picture, dating from summer 1970, shows 101/FC/8B. The vehicle has neither wheel step nor winch – but it does have a curious protrusion on the front hub that may have been intended as a miniaturized wheel step. LAND ROVER

one-ton trailer in comparison trials with the US-built M561 Gama Goat.

It was then returned to Rover, and on 17 July 1975 was sold into private ownership. It acquired the registration number OAA 224P and a chassis plate reading LR-101-FC-8. The vehicle still survives, in private ownership.

101/FC/8C (VXC 754K/ 03 SP 77; FVRDE Wing Number Not Known)

The 101 development team was originally permitted to build eight prototypes, but by late 1970 it became clear that they needed more. So a ninth vehicle was constructed from parts, including a spare chassis, in order to avoid putting in a request for more. Giving

The chassis frame of the 101 was a deliberately simple design. This was one of the later prototypes, 101/FC/8C, pictured during a weight-analysis exercise in the Rover engineering department during summer 1971. By this stage, the 3.5-litre V8 engine was firmly in the specification.
LAND ROVER

DESIGN AND PROTOTYPES

it the number 9 risked drawing the attention of management to the fact that the team had more prototypes than they were allowed, so Norman Busby 'disguised' it by having it numbered as 101/FC/8C.

The earliest reference to this vehicle in surviving job cards dates from early December 1970, when instructions were given to 'rebuild' it with a new chassis frame and 24V electrical system. 101/FC/8C was built with engine number 355-00031A, which was an early production Range Rover type. It was road-registered on 1 September 1971.

There were modifications to the rear body, affecting the weight, in October that year but the whole job card was not signed off (to indicate completion) until 20 June 1974! This delay was probably the result of oversight. The vehicle was also used for cold weather trials and was fitted with a winterization kit in January 1972. In March 1972, it was probably 101/FC/8C that was given additional reinforcement. *Land Rover Owner* magazine once reported that this vehicle was registered as 03 SP 77 during military trials, but this cannot be confirmed.

101/FC/9 (YXC 503K; FVRDE Wing Number Not Known)

Norman Busby obtained permission from managing director A. B. Smith on 8 September 1970 to build this final prototype. He issued a job card requesting its construction on 4 February 1971, and called for a vehicle to the '956' specification, which had been agreed for production (956-prefix chassis would have RHD and 12V electrical systems, but no production examples would be built before 1972). However, 101/FC/9 was to have a 24V electrical system, a plastic fuel tank, a Nokken winch and a powered-trailer drive system.

The powered-trailer drive system was robbed from 101/FC/8, and registration records show that

The registration plate identifies this vehicle as 101/FC/8C. It is substantially to production standard and has both taper-leaf springs and a Nokken winch, but curiously there is no wheel step fitted. The canvas hood lacks the small passenger's side window that would become standard. LAND ROVER

DESIGN AND PROTOTYPES 35

that 101/FC/9 had engine number 32, which was presumably 2158/32. That engine had earlier been used in a 1966 Rover P6B saloon prototype, number P6B/14, registered as JXC 815D.

The chassis was used for torsional tests in July 1971, and the complete vehicle was registered on 7 January 1972; the Solihull local registration book shows it as FC 101/9. In February 1972 it was rebuilt, using the heater from 101/FC/8A, in preparation for an unspecified trial. This may have been to test the standard heater against the winterization kit that had been fitted to 101/FC/8C. In May 1972, it was used for filming on Salisbury Plain; it was shown towing a 105mm gun for *International Defence Review*, and a 16mm film of the event was made for Rover.

Later in 1972, 101/FC/9 was put through a 10,000-mile test, but no further details of the vehicle are known.

RIGHT: **This is 101/FC/8C again, pictured in summer 1970. There are detail differences at the rear, where the lights are now mounted vertically and the configuration of the grab-rails and end plates for the inward-facing seats is different. This vehicle has the powered-trailer drive. The arrangements for ducting heat into the rear compartment have not yet been finalized.** TANK MUSEUM

BELOW: **YXC 503K was the final prototype, built during 1971. It is seen here towing a 105mm light gun; the barrel was traversed and was carried over the trail when the gun was being towed.** LAND ROVER

Technical Specifications, 6-Cylinder Prototypes

Engine

(1) Rover 6-Cylinder, Prototypes 101/FC/1 to 101/FC/6
Type: 3L7 6-cylinder petrol, with iron block and aluminium alloy cylinder head
2995cc (77.8 × 105mm)
Overhead inlet and side exhaust valves; chain-driven camshaft
Seven-bearing crankshaft
Compression ratio: 8.0:1
Single SU type HD6 (2in) carburettor
110bhp at 4,500rpm
150lb ft at 2,000rpm

(2) Ford Falcon 6-Cylinder, Prototype 101/FC/7 Only
XT Series 6-cylinder petrol, with iron block and aluminium alloy cylinder head
3620cc (93 × 88mm)
Overhead valves; camshaft chain-driven
Seven-bearing crankshaft
Compression ratio not known
Ford carburettor
135bhp at 4,400rpm (estimated)
170lb ft at 1,700rpm

Transmission

Four-speed primary gearbox, with single dry-plate clutch
    Ratios 3.37:1, 2.04:1, 1.37:1, 1.00:1, reverse 2.96:1 (with Rover engine)
    Ratios 4.609:1, 2.448:1, 1.505:1, 1:1, reverse 3.664:1 (with Ford Falcon engine)
Two-speed transfer gearbox, with 1.53:1 high ratio and 3.27:1 low ratio for Rover engine; 1.174:1 high ratio and 3.321:1 low ratio (for Ford Falcon engine)
Selectable four-wheel drive (with front axle disconnect)

Axle ratio

4.7:1 with Rover engine
5.57:1 with Ford Falcon engine

Suspension, Steering and Brakes:

Solid axles front and rear with semi-elliptic leaf springs
Recirculating ball steering with 19.6:1 ratio
Drum brakes on all four wheels, with servo assistance; 11 × 3in at the front, 11 × 2.25in at the rear; separate drum-type transmission parking brake; split hydraulic system with two Clayton Dewandre HSD 770 servos on 101/FC/7

Wheels and Tyres

6.50J × 16 steel disc wheels
9.00 ×16 tyres, Dunlop Trakgrip

Dimensions

Overall length:	169in (4,290mm)
Overall width:	72in (1,830mm)
Overall height:	86in (2,180mm)
Wheelbase:	101in (2,565mm)
Front track:	60in (1,520mm)
Rear track:	61in (1,550mm)
Ground clearance:	10in (254mm) under differentials

Unladen Weights

4,910lb (2,227kg) for 12V GS with winch
4,396lb (1,998kg) for 101/FC/7 with Ford Falcon engine*

---

* This figure may have been calculated differently from the figure for the Rover-engined prototypes, which was calculated by FVRDE.

CHAPTER THREE

PRE-PRODUCTION AND TRIALS

Standard Rover practice was to build a batch of vehicles to production specification before commencing volume production on the assembly lines. These vehicles would then be used to fine-tune the design and the assembly methods, and for various experimental tasks such as testing future design improvements and assessing in-service problems. They would be known as pre-production or pilot-production types, and would be numbered in the standard production chassis sequences. However, they would also be allocated engineering fleet numbers so that the project engineers could more easily keep track of them. These numbers followed on from the prototype sequence, so that the first vehicle became 101/FC/10.

In the case of the 101s, full production was still some way off by the time the MoD awarded the One-Tonne vehicle contract to Rover. Delays with the 105mm light gun now meant that it would not enter service until 1975, and so the new gun tractors were not required before then. This actually helped Rover out by resolving the issue of where to build the 101s, which were so different from any existing production Land Rover that they could not be assembled on the main Land Rover lines without causing major disruption. Over the summer of 1973, Rover ended production of its P5B (3.5-litre) Saloons and Coupés, which had been built in the assembly hall behind the main office block. So the 101 lines were set up here, ready to start volume production during 1974.

Meanwhile, and as so often happens in the military world, the user requirements were changing. The original plan had been to mount the 105mm light gun on a powered carriage, but that plan had given way to the idea of towing a powered ammunition limber trailer between the tractor vehicle and the gun itself. This made for an extremely long vehicle train, which certainly looked as if it would give manoeuvrability problems. The MoD seems to have been undecided about the value of the powered-trailer drive by 1972, and it is noteworthy that only three of the twenty-five pre-production 101 One-Tonnes built for the British military are known to have been fitted with the system. By the time full production began in 1975, the powered-trailer drive was no longer a requirement. Just a handful of production vehicles eventually had it.

THE PRE-PRODUCTION BATCH – 1972–73

A total of thirty-seven pre-production 101s were built in 1972–73, all assembled by hand at Solihull before the production line was ready. Of these, twenty-five were delivered to the MoD for user trials, ten were retained by Rover as demonstrators and engineering development vehicles, and two were built specially for a potential Canadian contract. All thirty-seven were painted in gloss bronze green, and all except the two Canadian trials vehicles were built as standard GS models.

Normal Rover practice would have been to build examples of each projected major type as pre-production vehicles, but in fact there were no pre-production 957-series (12V RHD Export) or 962-series models (24V RHD Export types). The thirty-seven pre-production vehicles were as follows:

03 SP 70 (956-00005A) demonstrates how much of the bodywork could be removed for airporting operations.
LAND ROVER

PRE-PRODUCTION AND TRIALS ■ 39

956-00001A to 956-00016A
   (12V RHD Home Market)   Total 16
959-00001A to 959-00002A
   (12V LHD Export)   Total 2
961-00001A to 961-00012A
   (24V RHD Home Market)   Total 12
964-00001A to 964-00007A
   (24V LHD Export)   Total 7

In the usual Rover fashion, these were not built in strict numerical order. The earliest one completed was actually 964-00001A, which entered the despatch department at Solihull on 21 April 1972. They also had a number of differences from the 'full production' vehicles because the detail design was still evolving. Even the pre-production models differed one from another, some, for example, having step rings on the front wheels and some being built without them. Like the final prototypes, the pre-production models had plain canvas tilts without the large side windows of the earliest vehicles. They resembled the production models quite closely, but the more obvious features of the pre-production 101s were:

- Front indicator and sidelights more closely spaced than on production models, with glass lenses
- No wheelarch spats
- Smaller door mirrors
- No lashing eyes on the first outrigger behind the front bumper
- Swinging spring shackles
- Vertically stacked rear lights

There is a more complete list of the differences between pre-production and volume-production 101 One-Tonne models in Chapter 4.

Some of the twenty-five vehicles allocated to the MoD were allocated registration numbers in the SP (Special Projects) series, while others were numbered in the FL series that had been allocated to the production 101s. Particularly interesting is that only three of them (956-00010A, 964-00004A and 964-00007A) were equipped with a powered-trailer drive. Quite clearly, the MoD had not made up its collective mind about the value of the powered trailer system

This profusion of instruction plates in the cab of pre-production 101 number 956-00002A includes the one reminding drivers to lock the centre differential before engaging the trailer drive. AUTHOR

by the time the first pre-production 101s were due to be built in 1972. A factor in this may have been that the new light gun itself had been delayed and would not be ready for another three years or so.

Of the other twelve pre-production 101s, two were built with truck cabs for trials in Canada (964-00002A and 964-00003A), and one went to Leykor in South Africa, presumably also for trials. Nine more remained with Rover at Solihull, as development and demonstration vehicles. Three were 956-series' vehicles (numbers 956-00002A, 956-00003A and 956-00006A) and just one was a 959-series' vehicle (959-00001A). There were two from the 961-series

(961-00001A and 961-00012A) and three from the 964-series (964-00001A, 964-00005A and 964-00007A). There seems to have been frequent exchanges of vehicles between MVEE and Rover, with the result that some vehicles had both civilian and military registration numbers.

IDENTITIES OF THE PRE-PRODUCTION VEHICLES

The pre-production 101s were shared between FVRDE and the Rover Engineering Department at Solihull. The lion's share (around two-thirds) went to FVRDE but the joint nature of the development programme ensured that many vehicles would shuttle back and forth between military users and Rover engineers. This has caused endless confusion about vehicle identities; even worse was when FVRDE decided they had too many LHD FFR vehicles and exchanged one with Rover for a RHD FFR type. Both vehicles carried the same military serial number and fleet identification code!

Rover's vehicles were mostly registered on civilian plates in the XC series allocated to the Solihull Registration Office. However, some of the Rover vehicles spent long periods in military hands, and when this occurred they were generally given military registration numbers. Some of the military examples were also reregistered with new military numbers if they were allocated to an engineering unit after serving in a field trials unit.

Within the engineering department, these pre-production vehicles were also allocated a fleet code. Some vehicles had a straightforward code in a series that continued from the prototype vehicles, which had ended with 101/FC/9. So the pre-production vehicles retained for engineering development became 101/FC/10, 101/FC/11 and so on. The two

The pre-production 101s were given a thorough work-out during user trials. This is 03 SP 72 on Exercise Hardfall in Norway during May 1975. Visible in the centre of the front panel, just below the windscreen, is the radiator header tank that was part of the winterization kit at this stage. MOD PUBLIC RELATIONS

PRE-PRODUCTION AND TRIALS ■ 41

HXC 677L demonstrates one of the problems with an unladen 101: the lightly loaded back-end tended to lift under braking, which was why a pressure limiter had to be added to the rear hydraulic line. This vehicle has a front NATO hitch, and was the Government Sales Department demonstrator. It had clearly been outside the UK not long before this picture was taken: the headlights still carry the beam deflectors required in countries where left-hand drive is the norm. LAND ROVER

vehicles for Canadian trials carried numbers beginning with PP. It looks as if Rover may have intended to use these numbers to identify all of the pre-production vehicles, but the only other one known to have carried such identification was the very first vehicle, which was known as PP1.

The twenty-five vehicles delivered to FVRDE were also given fleet identification codes beginning with a single letter P. Both the engineering fleet codes and the P codes were normally carried on a paper disc inserted into a standard tax disc holder and displayed on the windscreen of the vehicle. The P code was sometimes also displayed on the bridging plate, or chalked onto the vehicle's panelwork.

The table (pp. 42–45) shows what is known about the thirty-seven pre-production 101s. It is not yet complete, and research continues. Note that one RHD pre-production vehicle carried Royal Navy registration number 02 RN 37; this vehicle had a trailer drive and an early winterization kit, and carried fleet number 101/GS1, but it has not yet been possible to identify it further.

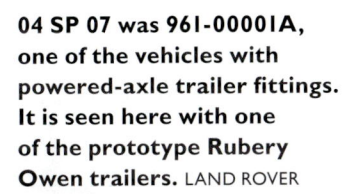

04 SP 07 was 961-00001A, one of the vehicles with powered-axle trailer fittings. It is seen here with one of the prototype Rubery Owen trailers. LAND ROVER

PRE-PRODUCTION AND TRIALS

Most of the pre-production vehicles had their rear lights stacked vertically, an arrangement that was discontinued for production. 03 SP 69 carries a 24V warning plate above the electrical socket on its rear cross-member. Just visible on the left-hand over-rider bracket is an unusual fairlead arrangement for the winch. This left-hand drive vehicle later became the **Vampire** prototype. TANK MUSEUM

Pre-Production 101s – 1972–73

Chassis no.	Fleet no.	FVRDE wing no.	Regn no.	Remarks
956-00001A	P1 or PP1	8301	03 SP 67; later 54 BT 06	Off build 1 August 1972. Fitted with rear seats.
956-00002A			KXC 808L	Off build 1 June 1972. Retained by Rover (Quality Control Department) and registered as KXC 808L on 29 June 1973. Engine number in Solihull records given as REL III. Fitted with trailer drive.
956-00003A			CXC 347K	Off build 1 June 1972. Retained by Rover (Engineering Department) and registered as CXC 347K on 20 June 1972.
956-00004A	P2	8302	03 SP 68	Off build 1 August 1972. Fitted with trailer drive.
956-00005A	P4	8316	03 SP 70; later 77 FL 29	Off build 30 August 1972. Built with winch and front towing hook; then to MVEE and used for a mock-up ambulance body. On 15 August 1977, it was re-registered as 77 FL 29.
956-00006A			Not known	Off build 25 September 1972. Retained by Rover and then shipped to Leykor, South Africa. Fate not known.
956-00007A	P7		60 FL 32	Off build 9 November 1972. To RSA (Royal School of Artillery) at Bulford Camp for light gun trials.

Pre-Production 101s – 1972–73

Chassis no.	Fleet no.	FVRDE wing no.	Regn no.	Remarks
956-00008A	P8		70 FL 83	Off build 6 December 1972. Fitted with winch and front towing hook. To RSA for light gun trials.
956-00009A	P12		60 FL 33	Off build 6 December 1972. To Warminster for special weapons trials.
956-00010A	P13		60 FL 34; later SXC 225M	Off build 12 December 1972. Initially to Warminster for special weapons trials. Registered by Rover on 8 March 1974 as SXC 225M. Engine 956-00008A; not originally built with trailer drive, but one was fitted for civilian demonstrations, probably in 1974.
956-00011A	P18		60 FL 35	Off build 12 February 1973. Possibly used for battery charger work.
956-00012A	P20		60 FL 36	Off build 12 March 1973. Possibly initially to REME.
956-00013A	P22		60 FL 37	Off build 8 February 1973. Possibly to MoD Quality Assurance Department initially. Not receipted at Ashchurch vehicle depot until January 1975.
956-00014A	P23		70 FL 84	Off build 12 March 1973. Fitted with winch and front towing hook. Probably to RSA for light gun trials.
956-00015A	P24		70 FL 85	Off build 22 March 1973. Fitted with winch and front towing hook. Probably to RSA for light gun trials.
956-00016A			HXC 677L; later FAA 105S	Off build 28 March 1973. Registered as HXC 677L on 26 March 1973. Retained by Rover (Government Sales Department). Used with Rubery Owen trailer no.3. Sold to Venturers Search and Rescue in late 1970s and acquired new civilian registration. To USA in mid-1980s.
959-00001A	101/FC/12	8392	EXC 870L	Off build 30 October 1972; registered on 03 November 1972. Engine number 956-00008A. Retained by Rover (Engineering Department).
959-00002A	P5	8327	03 SP 72; probably later 54 BT 11	Off build 4 October 1972. Fitted with a cold weather kit by MVEE; became the 'Norwegian trials' vehicle and was used on exercises in cold-climate conditions. It was intended to become 77 FL 28, but probably never did.

(continued overleaf)

Pre-Production 101s – 1972–73 (continued)

Chassis no.	Fleet no.	FVRDE wing no.	Regn no.	Remarks
961-00001A	101/GS5		EXC 384L; later 04 SP 07	Off build 20 September 1972. Retained by Rover (Government Sales Department) and registered as EXC 384L on 9 October 1972. Engine 961-00004A. Fitted with FVRDE-type powered-axle trailer drive and subsequently with the later Scottorn type. Used with powered trailer 15 EN 07. Sold to a UK private owner in July 1974, then to a private owner in Kenya, with number-plates EXC 384L.
961-00002A			Not known	Off build 13 April 1973. Retained by Rover (Government Sales Department).
961-00003A	P9		68 FL 39	Off build 3 November 1972. To RSA for light gun trials.
961-00004A	P10	8488	74 FL 11	Off build 12 December 1972. Fitted with winch, front towing hook and rear end dropped (towing) plate. To BAC for Rapier trials.
961-00005A	P11		74 FL 12	Off build 29 November 1972. Fitted with winch, front towing hook and rear end dropped (towing) plate. To RSA for trials with Rapier test equipment.
961-00006A	P14		68 FL 40	Off build 5 January 1973. Used for Beeswing missile trials.
961-00007A	P15		68 FL 41; later 04 SP 59	Off build 5 January 1973. Initially to Christchurch SRDE for FFR installation work. Chilwell records say it was re-numbered in January 1975, but other military records say it carried the SP registration between January 1973 and September 1978.
961-00008A	P16		68 FL 42	Off build 5 January 1973. Initially to Plessey for electronics work.
961-00009A	P17		68 FL 43	Off build 5 January 1973. Used for Beeswing missile trials.
961-00010A	P19		68 FL 44; later 05 SP 10	Off build 26 January 1973. Originally retained by Rover; exchanged in October 1973 for 964-00006A and took the military serial number originally allocated to that vehicle. Fitted with prototype signals body by Cammell Laird, probably in 1974.
961-00011A	P21		68 FL 45	Off build 26 January 1973. Receipted at Ashchurch vehicle depot January 1973. No further details.

Pre-Production 101s – 1972–73

Chassis no.	Fleet no.	FVRDE wing no.	Regn no.	Remarks
961-00012A	101/FC/13 P25		68 FL 46; NXC 629M	Off build 22 March 1973. Receipted at Ashchurch vehicle depot January 1973. Registered at Solihull 31 July 1973. Engine 956-00008A. Retained by Rover but then returned to MoD. Used for waterproofing trials at Instow.
964-00001A	101/FC/10		BXC 676K	Off build 24 April 1972; first pre-production vehicle completed. Retained by Rover (Engineering Department) and registered on 3 May 1972. Fleet number noted in Solihull local registration records; engine number 961-00001A. Fitted with metal cab in early 1974. Allocated to the Government Sales Department. Sold off on 17 June 1977. Survives in the Dunsfold Collection.
964-00002A	PP3		72-18756	Off build 1 June 1972. Canadian DND trials' vehicle no. 1, August 1972.
964-00003A	PP4		72-18757	Off build 1 June 1972. Canadian DND trials' vehicle no. 2, August 1972.
964-00004A	P3	8315	03 SP 69, later 54 BT 07	Off build 30 July 1972. Built with winch and front towing hook. Became Vampire prototype. Registration changed in January 1980.
964-00005A	P6	8326	03 SP 71, later 54 BT 08	Off build 4 October 1972. Registration changed in January 1980.
964-00006A	P19	8465	68 FL 44	Off build 4 October 1972. Issued to the School of Signals. Returned to Rover in October 1973, when it was exchanged for a RHD FFR vehicle (961-00010A).
964-00007A	101/FC/14		04 SP 08 (?); later JWK 748N	Off build 13 April 1973. Built with powered-axle trailer drive. Retained by Rover (Government Sales Department).

THE CANADIAN VEHICLES – 1972–73

Two of the pre-production 101s with left-hand drive were specially prepared for trials to meet a Canadian DND (Department of National Defence) requirement for a 1.25-ton payload vehicle. Preparation of the two 24V vehicles (numbers PP3 and PP4, 964-00002A and 964-00003A) began in June 1972 at Solihull, and both of them were specially fitted with an opening driver's side windscreen that was hinged at the top. The rear of the cab was also modified to create stowage space behind the seats, and the load bed was shortened at the front to create a gap between cab and body. Land Rover engineer Roger Crathorne recalls that their suspension was very much uprated to meet the Canadian requirement – perhaps to cope with a payload as great as two tonnes.

There were minor differences between the two trials vehicles. Number 964-00002A was fitted with

46 ■ PRE-PRODUCTION AND TRIALS

LEFT: **The two vehicles sent to Canada for trials had a special opening windscreen. This was 964-00002A, pictured in the workshop at Solihull.** BMIHT

BELOW: **Although the back body on the Canadian vehicles seems to have been standard, this gap between body and cab was not.** BMIHT

a winch, spare wheel behind the co-driver's seat and seats in the load-bed. The second vehicle, 964-00003A, had the spare wheel mounted behind the driver's seat. The two vehicles were delivered to the Canadian LETE (Land Engineering Test Establishment) at Orleans in Ontario in August 1972 and were given military identifications. 964-00002A was registered as 72-18756 and took LETE Control Number 1, while 964-00003A became 72-18757 with Control Number 2.

However, it appears that the Canadians were not impressed with the canvas-roofed cab. 'Subsequent to the initial acceptance,' reads one of the military tests report from the DND, 'a design change was introduced by the manufacturer whereby the soft-top cab was replaced with a hard-top cab.'

The logistics of this are not clear, but it is likely that the new cab was developed on another vehicle at Solihull and then sets of parts were sent to Canada to be fitted to the two trials vehicles. The new cabs simply added a roof (possibly made of GRP) with a ring-type hatch above the co-driver's seat, side extensions behind the doors and a rear panel with sliding window. The cab extension essentially filled in the space between the seats and the front of the load bed.

The trials themselves were conducted during 1973, the other contenders being the Volvo Laplander and pick-ups supplied by Chevrolet and Dodge. The earliest trials report was dated May 1973 and the

PRE-PRODUCTION AND TRIALS ■ 47

A special hard cab was designed after the Canadian vehicles had reached their destination, and was added to them before the trials began. These pictures show the two vehicles in semi-derelict condition in Canada during 1991.
ROBIN CRAIG

last one November. The 1.25-tonne Chevrolet pick-up won the contract, but Solihull chose not to take the two 101 trials vehicles back. So the DND sold them off, along with a large quantity of spares, to a scrap-car dealer. From here, they were bought by an Ottawa Valley enthusiast, but it appears that they were never returned to running condition and were semi-derelict by 1991.

PLANS FOR CIVILIAN VERSIONS – 1972–74

Even though the 101 had been drawn up to meet specific military requirements, Land Rover lost no time in investigating the potential for civilian-market sales as well. There had been forward-control Land Rovers available ever since 1962 and, although these

48

49

These 1972 renderings by Tony Poole in the styling department were intended to show how the 101 might be adapted to civilian market requirements.
LAND ROVER

were to end production in 1972 because of slow sales, some effort went into looking at the possibility of using the 101 to replace them.

It is not quite clear where the initiative for the investigation came from, or even how serious it was at that stage. One way or another, Tony Poole, who was then in charge of Land Rover work in the Rover styling department, made a series of six sketches that showed different potential civilian uses for the 101. Photographs of these sketches were filed in September 1972, so it is reasonable to suppose that they had been done shortly prior to that. Sadly, the photographs were in black and white; it is almost certain that the original sketches would have been in colour.

The sketches showed the 101 as an airfield crash-rescue tender, a mountain rescue ambulance, a gritting lorry with a snowplough, and fitted with a 'cherry-picker', a drilling rig and a backhoe digger. All of them were exactly the sort of thing that could have been developed and approved through what was then known as the Special Projects Department (and subsequently became Land Rover Special Vehicles).

MVEE did not hold back when testing the pre-production models. 03 SP 68 (956-0004A) was one of several that were overturned; it is little wonder that a sturdy rollover protection bar was requested for production vehicles. LAND ROVER

Testing at Solihull: this pre-production vehicle was deliberately rolled over to test the strength of the rollover bar mounted behind the cab. LAND ROVER

Although the 101 was certainly powerful enough to cope with the weight of ammunition limber and gun, the immense length of the combination would surely have made it a liability in some conditions. This was 04 SP 07 again. LAND ROVER

There is no doubt that a 'civilian' 101 would have been expensive. When *Contract Journal* reported on Land Rover's plans to commercialize the 101 in its issue of 18 April 1974, the projected likely price was quoted as around £4,000. This was at a time when a Range Rover cost under £3,000 in basic form and a 109in normal-control Land Rover cost under £1,700, inclusive in each case of all UK taxes.

There were thoughts about commercializing the powered-axle trailer with it but, thought *Contract Journal*, 'the powered trailer idea is unlikely to be cheap enough to appeal commercially'. The magazine was able to try a vehicle with powered-axle trailer (SXC 225M, which had been borrowed back from the MoD and re-registered) and found that it performed well, although the brakes 'were less efficient after deep water wading'. For most of the time, they wondered whether the third-axle drive on the trailer was really necessary, but 'the six-wheel drive did come into play...when the vehicle ploughed into a deep wet ditch. Four-wheel drive alone was not sufficient to climb out, but with six-wheel drive, that extra push from the trailer did the trick'.

Land Rover's view was 'that there could be worthwhile outlets in civil engineering and contracting, as well as for farming, forestry and special equipment'. The magazine agreed that the basic vehicle 'could prove very attractive for farming and forestry' and pointed out that the 101 was 'considerably cheaper than any vehicle with comparative *(sic)* performance', even though it was 'expensive as a one-ton pick-up'.

However, the 101 never became a civilian vehicle, because Land Rover discovered that its cab was too cramped to meet Construction and Use regulations in the UK. Presumably it could have been re-engineered to increase the space available, but the cost of doing so must have been considered too great to be recouped from the anticipated volume of sales.

MOD USER TRIALS – 1973–74

The twenty-five pre-production 101s allocated to the MoD were put through user trials in 1973 and 1974, and these trials identified a number of other areas for improvement. The vehicle's cornering stability came into question, and in 1974 engineer Roger Crathorne was attached temporarily to the development team to help out with developing an anti-roll bar installation on the front axle.

The final sign-off took place at the MVEE's Bagshot trials ground. The plan was for the 101 to be demonstrated in front of some very senior army officers: a vehicle from Land Rover's engineering fleet (in fact

NXC 629M) would be loaded with a test-rig load of a tonne to simulate heavy ammunition and would be driven at speed through a chicane course marked with cones. Roger Crathorne was chosen to drive the vehicle, and all went well on the practice run the day before the event. However, overnight the army cleaned and polished the vehicle, removed the test load, and did not secure it properly when they put it back. As a result, the load shifted during the demonstration run and the vehicle ended up on two wheels. Crathorne managed to get it back onto all four wheels again and stopped in front of the watching military brass; sales manager Terry Meakin persuaded them to think that had the whole thing had been intended to demonstrate the vehicle's stability and manoeuvrability!

LEFT: **At the final sign-off ceremony, pre-production vehicle NXC 629M nearly wiped out the attendees when the load shifted during a demonstration run around a chicane of cones. This picture was taken during a practice run before the event.** LAND ROVER

BELOW: **The Canadian vehicles were not the only ones to have hard cabs. In early 1974, pre-production model BXC 676K was also given one, of a different design.** BMIHT

CHAPTER FOUR

THE 101 IN PRODUCTION

A dedicated assembly line for the 101 was set up in the main production area at Solihull during 1974. It occupied space that had been unused since the Rover P5B saloons and coupés had gone out of production in summer 1973. This area was being used for Range Rover assembly at the time of writing in 2014, and is behind what is now known as Block 1, which still contains the offices of the company's directors.

The first four production vehicles were assembled on the line during November 1974 to test procedures, but volume production did not begin until January 1975. Production of 101s ended in May 1978 after a total of 2,667 had been built, giving an average annual production of around 760 vehicles. (There are more detailed statistics elsewhere in this chapter.) A stream of vehicles also returned to Solihull between 1977 and 1984 for rework of various kinds.

The vast majority of Land Rover 101s were built for the UK armed forces, but Land Rover lost no time in touting for overseas orders, and the first overseas deliveries of 101s were made during 1975, to Dubai and Luxembourg. About 10 per cent of all 101s were built for overseas armed forces (there is more detail about these in Chapter 7). In addition, a small number of vehicles were retained at Land Rover's Solihull factory, as engineering development or reference vehicles and as demonstrators, and did not enter military service.

Most 101s were built with the standard GS body and a full-length canvas tilt. However, a number were also completed as chassis-cabs to be fitted with special bodywork by specialist companies. The special-bodied British variants, such as the ambulance and signals-body types, are covered in Chapter 6.

By March 1973, plans had been drawn up to make the 101 available for CKD assembly overseas. However, by March 2014 the photographic record of them had disappeared from the Rover Company negatives held at the Heritage Motor Centre. The intended chassis sequences appear to have been 958 for RHD 12V models, 960 for LHD 12V models, 963 for RHD 24V models and 965 for LHD 24V models. No CKD production ever followed.

CHANGES FROM THE PRE-PRODUCTION VEHICLES

The user trials of the pre-production vehicles had shown up a number of areas for improvement, which were passed from FVRDE to Rover at Solihull as 'requests' during 1973 and 1974. In December 1974, Norman Busby drew up a list of the requested changes; in practice, the four 'production' vehicles that had already been assembled by hand had probably incorporated these changes. The four vehicles were used on the Trans-Sahara Expedition in 1974–75.

The list of changes that Norman Busby drew up was extensive, and consisted of the following:

BODY AND HOOD:
- Larger exterior mirrors
- Passenger's side window in hood curtain
- Rollover bar behind cab area
- Wheelarch spats added to outer sill channels
- Access holes in front floor for electrical items
- Rear body stiffeners

54 ■ THE 101 IN PRODUCTION

CAB:
- Safety belts added
- Interior mirror added
- Retention strap added to passenger's seat squab
- Larger lettering on transfer box warning plate
- Grab handle added
- Improve fixing for instrument panel

CHASSIS:
- Rear light and reflector positions changed (lights arranged horizontally rather than stacked vertically)
- Rear step added
- Wheel steps added
- Mud flaps front and rear added
- Goodyear tyres to replace Dunlop T24 type on trials' vehicles
- Extra chassis stiffeners to prevent cracking
- Redesigned battery carrier
- Revised helicopter lifting eyes on chassis
- Thinner wheel discs with 8swg instead of 7swg metal
- Modified rear bumperettes and fixings
- Paint winch cable rear near end as a warning
- Outrigger strengthened with patch plate
- Add anti-roll bar and modify chassis to suit
- Additional holes to fit Amphenol plug and cable assembly (i.e. multi-way electrical plug)
- Modified steering relay

ELECTRICAL:
- Horns changed
- Drawings required to show waterproofing of fuel pump and tank
- Add waterproof box for suppressor box and fuel pump
- Add waterproof connections for distributor and spark plug leads

POWERTRAIN:
- Raised air intake
- Relocate spark plug leads and add new vacuum pipe
- Drain plug for exhaust silencer

In practice, there were other changes between pre-production and production vehicles, too. The

This was the production rear lighting configuration. The step aided access to the rear of the vehicle when the tailgate was closed. AUTHOR

A standard NATO towing jaw was bolted to the rear cross-member on 101s for the UK armed forces. AUTHOR

The instrument panel was always basic and functional: speedometer and multi-instrument dial (with water, fuel and oil gauges) were flanked by various switches and warning lights. Nearest the camera is the military lighting switch, and under a protective shield next to it is the infra-red lighting switch. AUTHOR

The gear lever was angled backwards and always came with a zipped protective leather gaiter. AUTHOR

aluminium housing of the Nokken winch was modified, and where the pre-production fuel tanks had their strengthening swages pressed out, the production type had the swages pressed in.

THE LONG TEST-DRIVE

The first four production vehicles were assembled on the new production line in November 1974, but they were the only 101s built that year. These four vehicles were allocated chassis numbers 956-00021A to 956-00024A. These did not follow on directly from the pre-production models that had already been assembled, as those had ended with chassis number 956-00016A. Numbers 17 to 20 were not built until 1975, so presumably the intention was to ensure that these four vehicles had more memorable identification numbers.

There was good reason why they should. Land Rover had agreed to supply the four vehicles for a

THE 101 IN PRODUCTION

Joint Services' Expedition, which planned to make the first crossing of the Sahara Desert from west to east; the desert had been crossed in the opposite direction before, but never this way. Part of the plan was to use the expedition as a development exercise in connection with the powered trailer, which at that stage still figured in military thinking. So three of the vehicles were fitted with the powered-trailer drive, and two prototype powered-axle trailers were also prepared to go with them. The four vehicles were given the military registration numbers 60 FL38, 76 FL 64, 76 FL 65 and 76 FL 66, the one without the trailer drive being 60 FL 38. The two trailers also carried their own registration numbers, these being 01 SP 43 and (probably) 02 SP 04.

The expedition was led by explorer Tom Sheppard, then still serving as a squadron leader in the RAF, and among its tasks were to be geophysical, geological and zoological studies on behalf of a number of scientific bodies, which included the Royal Geographical Society. The vehicles went from the assembly line to the experimental department at Solihull for detail preparation to meet Tom Sheppard's wishes. They were worked on by experimental department fitters, and also by members of the expedition team: Sheppard wanted to ensure that everybody involved was fully familiar with the vehicles before they set off. The four 101s then went to RAF Uxbridge in November 1974 for final preparation,

Clear here are: the ducting above the engine cover that took heat to the rear compartment; the electrical shunt box used on 24V vehicles; the sturdy rollover bar behind the cab; the Perspex window in the passenger's side of the canvas hood (this is a LHD vehicle); and the plain finish of the standard production seats.
AUTHOR

Expedition leader Tom Sheppard photographed two of the Trans-Sahara vehicles being carefully finished in the experimental department at Solihull. The fitters working on the nearer one are Brian Finch and Ron Yardley.
LAND ROVER/TOM SHEPPARD

ABOVE: **The four Trans-Sahara vehicles are seen here lined up at Solihull in 1974, before delivery to RAF Northolt. With them are members of the expedition team.** LAND ROVER

RIGHT: **Many years later, the first production vehicle, still wearing its expedition insignia on the door, was pictured with Tom Sheppard during a Land Rover media event.** LAND ROVER

and the expedition set off on 15 January 1975 from Westminster Bridge in London.

By 26 January, the vehicles were in Dakar for the start of the Sahara crossing. They were heavily laden for the desert, with loads averaging 2,553lb (1,158kg) each, or more than 43lb (19.5kg) over their planned load capacity. Each member of the team was allowed personal items weighing no more than 50lb (22.7kg), and spread among the vehicles and trailers were 900 gallons (4,091ltr) of petrol and 340 gallons (1,546ltr) of water.

To the satisfaction of Land Rover engineers and the expedition team alike, there were no major failures during the 100 days of the expedition. The biggest problems were overheating engines. There were numerous oil leaks from such items as the gearbox, transfer box, bellhousing drain hole, steering box and swivel-pin housing seals; black plastic swarf was found in the carburettors and a multitude of bolts worked loose. Yet over the 7,494 miles (11,990km) they covered, the vehicles needed only one oil change, and that was in Nigeria. The powered-axle system returned a less impressive performance, however. The receiver hitch in the cross-member of the towing vehicles overheated badly, and one trailer lost a wheel in the harsh conditions of the desert.

PRODUCTION MODELS

Land Rover built most of the 101 One-Tonnes with the standard GS body, but it also built a sizeable number of 101s in chassis-cab form. These were mainly intended to meet UK MoD demand for box-bodied signals and ambulance vehicles, whose bodies were to be constructed elsewhere.

The standard GS (General Service) body was a simple dropside type, and an unusual characteristic of its construction was the use of Avdelok rivets with counter-sunk heads; these were put in place with a special rivet gun. The cab of the GS vehicles had no roof. To the metal bodywork was added a full-length canvas hood that covered both cab and load area. These hoods (or 'tilts') always had a Perspex window in the passenger's side of the curtain linking back body to cab, to give the driver a better view of the kerb side of the vehicle. As a result, the canvas hoods for RHD and LHD vehicles differed.

Vehicles were built at Solihull with both RHD and LHD, and with either 12V or 24V electrical systems to suit their anticipated use. Some vehicles were fitted with winches – the Nokken R1B type. Only a handful of production vehicles were built with the

964-00461A was built towards the end of production in 1978 but was never delivered to the military. Retained at Solihull, it later passed into the collection of vehicles retained by the Land Rover Experience at Solihull. It typifies the way the vehicles looked when they left the factory. AUTHOR

powered-trailer drive, which had been removed from the UK MoD requirement before volume production began.

All vehicles for the British armed forces left the Solihull factory with deep bronze green gloss paint, and their canvas hoods were in khaki drab. Some vehicles built for overseas users were painted in different colours and had different hood colours: the hood colours available were listed as drab, stone and khaki.

A word of explanation is necessary about the chassis-type descriptions. Rover traditionally assumed that all LHD vehicles they built were intended for export, but of course many LHD 101s were not exported in the traditional sense but were allocated to units of the British armed forces serving overseas in countries where LHD was the norm. The RHD export sequences were traditionally used not only for RHD vehicles destined for export, but also for those that varied from the home market specification in some significant way and were not intended for export abroad.

The 24V vehicles, also known as FFR (fitted for radio) types, were built with standard military-pattern 90A three-phase alternators, and had their distributors, coils, HT leads and spark plugs all waterproofed and screened. In addition, electrical equipment was properly earthed with braid to suppress interference. They were normally fitted with hand throttles so that the engines could be run at an appropriate speed while the vehicles were stationary in order to charge the batteries.

Towards the end of planned production, the UK MoD asked the Rover Company if it would consider building a further 2,000 vehicles over a ten-year period. However, as Norman Busby remembered it many years later, 'the thinking was, "If we can't make more than twenty a week, it's not economical"...

An earlier vehicle retained at Solihull as an engineering reference vehicle was 956-00168A. Unusually, this was a production model built with a powered-trailer drive. It now belongs to the collection of the Heritage Motor Centre at Gaydon. The canvas hood with its paired side windows was not typical of production models, and in this case is a LHD type on a RHD vehicle. LAND ROVER

Our clever sales people said, "Oh, no. If you'll have the lot, OK, but not over ten years." So the British Army didn't order its 2,000'. Building 2,000 over a period of 10 years would have demanded very low-volume production of about four vehicles a week, and even if other potential orders in the pipeline from the Malayans and the Egyptians had been confirmed, the numbers would not have been viable. So there was no extension of the 101 contract, and production came to an end in April 1978.

Pioneer tools were kept in brackets on the front panel below the windscreen.
AUTHOR

The last 101 of all appears to have been 962-00092A, which left the assembly line on 18 May 1978 and was delivered to BAC at Stevenage. The last home market model was 956-01322A, which left the line on 13 April 1978 and was delivered to the Ministry of Defence.

The red-painted filler cap reminded drivers that this vehicle used petrol; the British military used yellow caps on vehicles with diesel engines. AUTHOR

The production models had angled locker lids at the rear of the body; prototype versions had been rectangular. This one was used to stow the wheel-changing equipment. AUTHOR

The 24V vehicles all carried a warning plate above the socket for the trailer electrics. Also visible here are the two leaves of the taper-leaf springs. AUTHOR

> ### THE NOKKEN WINCH
>
> The Nokken R1BJ winch that was used on around 20 per cent of the 101 One Tonne models s an unusual type to be found on a vehicle. Normally used in forestry, it was manufactured by Hordaland Mek in Norway.
>
> A key advantage is its side mounting on the chassis frame, which makes it readily accessible and also allows the winch cable to be fed to either the front or the rear of the vehicle. The 65m (212ft) cable is twice as long as that on the typical vehicle-borne self-recovery winch, and the load on the cable is constant because the Nokken is a capstan-type winch; on drum winches the load varies with the amount of cable still on the drum.
>
> The winch is driven from the centre power take-off on the transmission, and its speed can therefore be varied to suit requirements by adjusting the engine speed.

CHANGES DURING PRODUCTION

There were no major changes to the 101 build specification between the first vehicles in January 1975 and the last one built in May 1978. So all 101s of all types always had the same A suffix to their chassis numbers. (Land Rover used these letter suffixes to indicate specification changes that affected the servicing of the vehicle.)

The British Army's 1978 report on the 101 in service is summarized in Chapter 5, and it led to a number of modifications that were designed to increase reliability. These were made on production vehicles, and they may account for the rework programme that was begun in mid-1977 (by which time Rover may well have had a preview of the report's findings) and which led to unspecified production modifications thereafter.

The production changes were as follows:

- An additional check ensured that the accelerator cable was grease-packed, as originally planned; it had not been on some earlier vehicles
- Another additional check ensured that steering stops and drag links were correctly set up on production; failure to do this had put the steering joints under extra stress and caused premature wear
- A modified accelerator relay lever was designed because the pinch bolt on the original pedal relay could work loose
- Changes were made to the synchromesh, which had failed on some vehicles in service
- The High–Low range selector forks in the transfer gearbox were modified because they had distorted on some early vehicles
- A modification was introduced to prevent the gear lever housing bracket from working loose
- The hub seals were redesigned because the originals tended to leak

Unfortunately, one other problem identified in the report could not be so readily eradicated. This was that the high propshaft angles inherent in the design led to UJ failures, to torn gaiters and to loose flange bolts.

'PATTERN' GS BODIES

A number of unissued chassis-cab vehicles were withdrawn from storage in approximately 1982 and may have been among those returned to Land Rover at Solihull for re-work before they were delivered to Marshall's in Cambridge. The plan was for them to be converted to ambulances but the full quantity of ambulance conversions was never made. A total of sixty-nine chassis, presumably already stripped down as necessary in preparation for the ambulance conversion, was left over.

So Marshall's were asked to build new bodies for them to GS pattern. These bodies had a number of differences from the original type. The most obvious difference is in the grab handles at the rear, which are angular instead of curved. The rear cross-member is straight above the tow hitch, where the originals had a curved section, and there are also differences in the body fastenings. It is probable, but not certain, that all sixty-nine chassis left over from the ambulance contract were given these 'pattern' GS bodies. These sixty-nine vehicles probably all became Rapier tractors for RAF 6 Wing (see Chapter 6). However, there were seventy such vehicles; the discrepancy of one has not yet been explained.

One for the rivet-counters! The standard production body sides (*left*) used spot welds; the Marshall's 'pattern' bodies had visible rivets (*right*). 101FCC&R

101 CHASSIS NUMBERS

These are based on the Land Rover despatch records, which are now held by the British Motor Industry Heritage Trust at the Heritage Motor Centre in Gaydon, Warwickshire (01926 641188).

Four series were reserved for CKD vehicles, but none was ever used. The series were prefixed 958- (CKD, RHD, 12V), 960- (CKD, LHD 12V), 963- (CKD, RHD, 24V) and 965- (CKD, LHD, 24V).

956-00001A to 956-01322A	Home market, 12V (1972–78)
957-00001A to 957-00028A	RHD export, 12V (1976–77)
959-00001A to 959-00583A	LHD, 12V (1972 and 1975–77)
961-00001A to 961-00170A	Home market, 24V (1972–73 and 1975–77)
962-00001A to 962-00092A	RHD export, 24V (1977–78)
964-00001A to 964-00474A	LHD, 24V (1972–73 and 1975–78)

NOTE 1: The chassis numbering system of the 101s did not change in 1975 to reflect the system used from that date on most other Land Rover models.

NOTE 2: These figures give a build total of 2,669 vehicles, but it appears that two vehicles were not built (959-00055A and 959-00411A) and that one of the vehicles intended for Luxembourg (964-00395A) was burnt out at the factory. The build total would therefore be 2,667 vehicles, of which 2,666 were actually delivered.

64 ■ THE 101 IN PRODUCTION

The pictures that follow are from stripdown and weight-analysis work done at Solihull, and give a good idea of what went into a 101 One-Tonne. These two show the chassis. The number-plate is one of those used on chassis-cabs while in storage in the army depots, and the stripdown of that vehicle may have been done in preparation for the ambulance contract.
LAND ROVER

ABOVE: **This is the cab structure, seen here sitting behind the stripped chassis.** LAND ROVER

RIGHT: **Axles and propshafts were laid out for inspection here.** LAND ROVER

66 ■ THE 101 IN PRODUCTION

LEFT: **This is the back body, with the rollover bar somewhat illogically positioned behind it.** LAND ROVER

BELOW: **These two pictures show the V8 engine with the gear lever attached, and also stripped to component form. In practice, the gear lever faced forwards, not backwards.** LAND ROVER

THE 101 IN PRODUCTION ■ 67

These are the components of the LT95 gearbox with its integral transfer box.
LAND ROVER

PRODUCTION TOTALS

Although the production chassis sequences suggest that 2,669 examples of the 101 were built, only 2,666 actually left the factory. The theoretical and actual production figures are shown in the table.

There were more 101s with RHD than with LHD – 1,612 as against 1,054. There were also more with 12V electrical systems than with 24V systems – 1,931 as against 735.

Theoretical and Actual Production Figures

Chassis sequence	Theoretical total	Delivered total
956	1,322	1,322
957	28	28
959	583	581
961	170	70
962	92	92
964	474	473
	(2,669)	**2,666**

THE 956-SERIES VEHICLES (12V RHD HOME MARKET)

How Many?
The 956 series was the most numerous type of 101 One-Tonne. Chassis numbers for these vehicles ran to 1322A, and all appear to have been used. The first sixteen were built in 1972–73 as pre-production models. Production began in late 1974 with the four vehicles for the Trans-Sahara Expedition, but there were no more vehicles built until early 1975. So 1,302 more production models were built between then and the end of production in May 1978.

End-Users
Most of the production vehicles were delivered to one or other branch of the British armed forces. However, two were retained at Solihull, and these were numbers 956-00167A and 956-00168A. Number 168A was fitted with a powered-trailer drive and became an engineering reference vehicle, being kept in unmodified condition as a sample of the standard product. Many years later it was refurbished and now belongs to the Heritage Motor Centre at Gaydon. It was never registered when new and remains unregistered to this day.

One of the 956-series vehicles (956-00849A) went to the Royal Ordnance Factory at Nottingham, where it was used for test-towing the light gun that was made there. It was still there as late as 2000.

There is no record of what happened to 956-01305A.

Later Conversions
Military records show that 1,060 of these vehicles remained standard GS trucks. Just one (65 FL 58) was given a signals body, and 261 were given ambulance bodies, as explained in Chapter 6. Of these, 259 had been built as chassis-cabs and remained in storage until the early 1980s.

Two of the 956-series 101s were converted to recovery trucks by units in the field.

The production records show that thirteen vehicles were re-worked at Solihull between 17 and 25 May 1977. The earliest numerically of these was 956-00828A and the latest was 956-00978A. It appears that all later vehicles were built with the changes that characterized this re-work.

There was also an out-of-gate re-work programme for some 956-series chassis, suggesting that Solihull modified some chassis after delivery to the MoD depots. Two vehicles (956-00986A and 956-01281A) were subjected to it in January 1978, and it may be that quality checks had not picked up faults that were subsequently rectified on military premises. Several years later, and after production had ended, a further thirty-five or thirty-six (the quantity is disputed) of the 956-series chassis received out-of-gate re-work between June 1983 and March 1984, before becoming Rapier tractors with RAF 6 Wing (see Chapter 6).

Three 956-series 101s were converted in service to FRT recovery vehicles (see Chapter 6).

THE 957-SERIES VEHICLES (12V RHD EXPORT)

How Many?
There were just twenty-eight of these vehicles, all built in 1976–77. All of them were delivered in deep bronze green gloss paint with standard canvas hoods.

End-Users
The first six examples went to Uganda, six more went to Brunei in mid-1977 and the last sixteen went to Kenya in mid-1977. All except the Ugandan examples were re-worked before despatch. (There are further details about these vehicles in Chapter 7.)

THE 959-SERIES VEHICLES (12V LHD EXPORT)

How Many?
The first two in this series were built as pre-production models in 1972. No more were then built until 1975, and production continued into 1977. The chassis numbers ran to 583A, but two vehicles (959-00055A and 959-00411A) were not built. This gives a build total of 581 for the 959-series.

Production modifications seem to have been introduced towards the end of 1976, although it is not clear what they were. The changes were standardized in production at chassis number 959-00410A, and a number of earlier vehicles were re-worked to incorporate the changes. Those known include three for the UK MoD (959-00401A, 404A and 405A),

THE 101 IN PRODUCTION 69

FUNERAL CARRIAGE

62 FL 66 (956-00293A) was modified to carry the coffin of Lord Louis Mountbatten at his state funeral in 1979. The vehicle had been in service with the Life Guards at Windsor. Mountbatten was assassinated on 27 August 1979 by the Provisional IRA, who blew up his boat off the coast of County Sligo, Ireland, near his family holiday home at Classiebawn Castle. Two of Mountbatten's relations and a 15-year old local boy were also killed.

62 FL 66 was used to carry the coffin of Lord Mountbatten. It is pictured here (*below*) on military premises after its special preparation, and at London's Waterloo station during the funeral procession (*bottom*). *TOP IMAGE:* MoD, *BOTTOM IMAGE:* DOMINIC CLINTON

twenty-one of the larger batch of vehicles delivered to Oman and one other originally intended for Oman but never delivered.

End-Users

As usual, the majority of these vehicles went to the UK MoD, the actual total being 494. The other eighty-seven vehicles were made up of twenty-two for the UAE in 1976, thirteen for Dubai in 1975–76, forty-two for Oman in 1976–77, three for Ras al Khaimah in the UAE in 1977, one (959-00282A) sold on the civilian market in the UK, apparently as a failed export order, and six (959-000135A to 139A) that went to British Aerospace but have not been traced. These might have included the Swingfire vehicles for Egypt (see Chapter 7).

The UAE order appears to have been originally for thirty-two vehicles, and these (numbered 959-00103A to 134A) were all delivered to BAC at Stevenage between April and June 1976. They were painted matt light stone (which has Land Rover colour code LRC 228). However, ten of them later appeared in British military service with 1983-issue registration numbers 42 KB 59 to 42 KB 68.

Of the UK MoD vehicles, twenty-one became tractors for the Blindfire tracking radar used for the Rapier missile system, and twelve were completed as Rapier test vehicles. A further quantity went to the RAF.

Later Conversions

A total of 192 unissued vehicles in the 959 series were withdrawn from storage in the early 1980s for conversion to ambulances; 172 were for the army and the remaining twenty for the RAF.

One 959-series 101 became an FRT recovery vehicle.

THE 961-SERIES VEHICLES (24V RHD HOME MARKET)

How Many?

There were 170 of the 961-series vehicles. Of these, the first twelve were built as pre-production models in 1972–73. The remaining vehicles were built in 1975–77.

End-Users

With the exception of three that remained in the Engineering Department at Solihull, all of these vehicles went to the UK MoD. The three engineering vehicles were numbers 961-00001A and 961-00012 (which were pre-production examples) and one production vehicle, 961-00019A. All three, nevertheless, carried military registrations when used for various military trials, the numbers being 03 SP 78, 68 FL 46 and 05 SP 11, respectively.

There were sixty-one vehicles fitted with Nokken winches, front pintles and rear drop-plates with towing pintles. Of these, forty-seven were completed as Rapier missile tractors, thirty-eight for the army and the remaining nine for the RAF.

The vehicles without winches were used in a variety of roles by REME units, and as command posts, FACE carriers, Milan carriers and mortar carriers.

The final 961-series vehicle, 961-00170A, was delivered to Farnborough in July 1977, but its duties there remain unknown. A reasonable guess is that it may have been used by the Army Personnel Research Establishment; if so, it would have had a civilian registration number, which has not been traced.

Later Conversions

Military records suggest that thirty-three of the 961-series chassis were converted to signals bodied vehicles by Marshall's. However, not all were equipped with radio gear and some were used for other purposes, such as electronic repair workshops, command posts and, later, BDS (Biological Detection Systems) vehicles.

Two 961-series 101s were converted in service to Challenger field workshop support vehicles.

THE 962-SERIES VEHICLES (24V RHD EXPORT)

How Many?

There were just ninety-two of these vehicles, all built towards the end of production in 1977–78.

End-Users

The first thirty-four vehicles went to Brunei, and a further fifty went to Australia as Rapier tractors. The remaining eight vehicles went to British Aerospace but their end-users have not been traced.

THE 964-SERIES VEHICLES (24V LHD EXPORT)

How Many?
There were 473 vehicles in the 964 series. Chassis numbers actually ran to 474, but one vehicle (964-00395A) was burnt out at Solihull. Seven vehicles were built in 1972–73 as pre-production examples. Full production then began in 1975 and continued through to 1978.

End-Users
Most of the 964-series vehicles were delivered to the UK MoD, but two were retained by Rover. These were 964-00001A, which belonged to the Government Sales Department, and 964-00038A, which belonged to the engineering department. Two more, 964-00002A and -00003A, went to Canada as trials vehicles (see Chapter 3). There were also several overseas deliveries.

The end-users of some of these vehicles have still not been identified, nearly 40 years after they were built. The largest number of 964-series vehicles sold overseas went to Luxembourg, which took sixty-one vehicles in 1975–76. It is probable that Egypt took three in 1976–77. However, thirty-six that were delivered to British Aerospace in Stevenage, probably for conversion to Rapier tractors, have not been traced.

Of the UK MoD vehicles, large numbers were completed as tractor units for the Rapier missile system. Many served with squadrons of 4 Wing RAF Regiment in Germany.

A total of eighteen late-production vehicles stood unused at the Ashchurch vehicle depot from 1980; these were registered as 75 GJ 17 to 75 GJ 34. In 1986, eleven of them (75 GJ 17 to 75 GJ 27) were issued to RAF Rapier Squadrons in Germany, but the other seven remained in storage at Ashchurch until early 1994. They were then sold off at Aston Down, still in mint condition and in their original gloss bronze green paint, although spares were removed first.

Later Conversions
In the early 1980s, a number of unissued 964-series vehicles were withdrawn from storage and were converted as elements in the Vampire electronic warfare system. Probably eighteen became Vampire vehicles, and a further twenty-five were converted to house the Intercept Complex. (There are more details of the Vampire system in Chapter 6.)

WHAT'S IN A NAME?

The Land Rover 101 Forward Control is known by a variety of names, as follows:

- 101 Forward Control. This was the name applied by the manufacturers, and remains in common use among enthusiasts.
- One-Tonne. This name was often used by British soldiers. It is usually pronounced as 'one-tunny' to emphasize that the vehicle's payload was, initially at least, a metric tonne (2,205lb) and not an Imperial ton (2,240lb). The vehicle was in fact always capable of carrying an Imperial ton. It was then rerated to 1.25 tons in order to carry the box and ambulance bodies, and the weight classification plates on later vehicles (from 1977 approximately) were amended to show a 3,652lb GVW instead of the earlier 3,143lb figure.
- FV 19000. This was the official military vehicle classification (FV stands for Fighting Vehicle) and was very rarely used.

Technical Specifications, 101 One-Tonne Production Models

Engine
V8 petrol, with aluminium alloy block and cylinder head
3528cc (88.9mm × 71.1mm)
Overhead valves; chain-driven camshaft
Five-bearing crankshaft
Compression ratio: 8.5:1
Two Zenith 175 CD 2S carburettors
    128bhp (gross) at 5,000rpm
    120bhp (nett) at 5,000rpm for 12V models
    116bhp (nett) at 5,000rpm for 24V models
    185lb ft (gross) at 2,500rpm
    176lb ft (nett) at 2,500rpm for 12V models
    170lb ft (nett) at 2,500rpm for 24V models

Transmission
Four-speed type LT95 primary gearbox, with single dry-plate clutch; ratios 4.069:1, 2.448:1, 1.505:1, 1.000:1, reverse 3.665:1
Integral two-speed transfer gearbox, with 1.174:1 high ratio and 3.321:1 low ratio
Permanent four-wheel drive with vacuum-lockable centre differential.

Axle Ratio
5.571:1

Suspension, Steering and Brakes
Solid axles front and rear, with semi-elliptic taper-leaf springs all round, 3in (76mm) wide, mounted above the axles. Front anti-roll bar.
Recirculating-ball steering with 23.3:1 ratio
Drum brakes on all four wheels, with vacuum servo and pressure-apportioning valve in the rear hydraulic line; 11 × 3in at the front, 11 × 2.25in at the rear; separate drum-type transmission parking brake.

Wheels and Tyres
6.50J × 16 steel disc wheels
9.00×16 tyres, Dunlop Trakgrip

Electrical System
12V with 16ACR 34A alternator or
    24V with AC90 90A alternator

Dimensions
Overall length:	170.5in (4,330mm) fully equipped
	166in (4,217mm), stripped
Overall width:	72.5in (1,842mm)
Overall height:	90in (2,283mm) over canvas tilt
	84in (2,138mm) to top of windscreen
Wheelbase:	101in (2,565mm)
Front track:	60in (1,524mm)
Rear track:	61in (1,549mm)
Ground clearance:	10in (254mm) under differentials

Unladen Weights
(These weights are for vehicles with coolant, oil, and 24 gallons/109ltr of fuel)
4,242lb (1,924kg) for 12V GS soft-top models
3,500lb (1,580kg) when fully stripped for airporting operations
4,259lb (1,940kg) for 24V GS soft-top models

CHAPTER FIVE

THE 101 IN BRITISH MILITARY SERVICE

All three branches of the British armed forces took examples of the 101. The army was by far the largest user, the RAF came a poor second and the Royal Navy probably had only two – although large numbers were used by Royal Marines units.

Exactly how many 101s actually entered British military service is nevertheless a matter of some contention. There are gaps and contradictions in some of the crucial records, and many vehicles were transferred between services, sometimes acquiring a new registration serial number in the process. Others had two or even three different registration numbers. So the figure of 2,380 calculated for this book should be considered a best guess. It includes the pre-production vehicles but not the engineering prototypes that are discussed in Chapter 2.

In practice, the requirement for the 101s had changed quite dramatically by the time they entered service. The original plan to use them as tractors for the 105mm light gun changed when the powered gun carriage was deleted from the gun's specification. As Chapter 2 explains, the powered-trailer drive was re-assigned to work with a simple trailer that served as a gun limber, but then problems with that persuaded the army to delete the trailer drive from the 101's specification.

Nevertheless, the 101s were used as gun tractors, as originally planned and in the planned quantities. However, a problem was that the front-line requirement of the British Army of the Rhine (BAOR) in Germany was changing. They now wanted more heavy armour instead of light artillery towed by

Wearing camouflage netting, this LHD 101 was pictured on the Hohne ranges in Germany in 1978.
PUBLIC RELATIONS, HQ
1ST ARMOURED DIVISION

70 FL 34 (959-00318A) was pictured serving with the 9th/12th Royal Lancers in Germany in 1979. PUBLIC RELATIONS, BAOR

BRITISH MILITARY CONTRACTS FOR THE 101 ONE-TONNE

WV 9615 – 3 December 1971
This was the first contract for production 101s and it covered vehicles delivered in the registration block 60 FL 32 to 77 FL 05. Specifically, the vehicles called for, all to Specification LV82, were as follows:

GS 1-Tonne FFR, with winch:	80 × LHD
	49 × RHD
GS 1-Tonne Cargo, with winch:	169 × LHD
GS 1-Tonne FFR:	351 × RHD
	186 × LHD
GS 1-Tonne Cargo:	684 × RHD
	139 × LHD
Total:	**1,658**

The contract was subsequently amended to call for additional vehicles. The final delivery total under this contract was 1,748 vehicles.

WV 12074 – 26 March 1976
This was the contract for the chassis-cabs that were originally intended to be bodied as ambulances. It was amended to reflect the reduction in 101 ambulances for the RAF and their subsequent transfer to the army. The ambulances were registered between 71 GJ 51 and 75 GJ 34, and from 75 GJ 50 to 76 GJ 02; the last (76 GJ 02) did not enter service until February 1984.

FVE21A/27 – Early 1977
This was the contract with Marshall's for the signals bodies.

FVE21A/156 – December 1979
This was the contract with Marshall's for the ambulance bodies. The contract was amended considerably; many of the planned RAF orders were either cancelled or turned over to the army before deliveries were made. A total of 519 were planned although only 450 were actually built and 69 were cancelled. The 69 'pattern' GS bodies from Marshall's were built under an amendment to this contract.

FVE21A/230
This contract covered RAF vehicles 00 AM 72 to 01 AM 41.

FVE22A/5
This contract covered five vehicles for the RAF, which became 83 AM 76 to 83 AM 80.

THE 101 IN BRITISH MILITARY SERVICE

soft-skin trucks. So there were multiple amendments to the two main MoD contracts for the 101s, and these reflected the changing requirements. In particular, far fewer 24V FFR vehicles were eventually delivered than had been ordered in the beginning, although the reason for this is not clear. One possibility is that signals regiments were concerned about the possibility of theft from canvas-topped vehicles, and wanted hard-body vehicles for greater security.

Determined to find uses for the quantity of vehicles it had ordered, the MoD decided to use a large number as towing and support vehicles for the Rapier missile system: 109in Land Rovers had been planned for this role, but they would have been very overloaded. It decided to reconfigure another quantity as hard-body types, and Chapter 6 explores the multiple uses of those. The use of the 101 as a battlefield ambulance had been in the plan from the beginning, but even then the contract was amended several times as vehicles were swapped between army and RAF, and so required different configurations.

DELIVERIES

Deliveries from the Rover works at Solihull to the two major military vehicle depots at the UK began in early 1975 and continued through to 1978. However, not every 101 delivered was immediately issued to a unit. Several hundred were placed in storage in anticipation of their later use as ambulance and signals' vehicles. Many of the chassis planned for ambulances – perhaps all – were given temporary military serial numbers in the CC series that was reserved for chassis-cab types. The earliest of these may have been 72 CC 63.

The first deliveries to the army were made at the start of 1975. There were fifty-six of the 956-series vehicles (12V RHD types), four in the 961-series (24V RHD types) and four more in the 964-series (24V vehicles with LHD). Two of the 961-series and all four of the 964-series vehicles are recorded as Rapier missile tractors, and had presumably passed through BAC at Stevenage for appropriate work after basic assembly had been completed at Solihull. (For more detail on the Rapier tractors, see Chapter 6.)

All these first deliveries were made to the army's central vehicle depot at Ashchurch in Gloucestershire, from where they would subsequently have been allocated to military units. Later deliveries were split between Ashchurch and the second major vehicle depot at Hilton in Derbyshire. All vehicles were given a receipt voucher (prefixed by ASH or ASC for Ashchurch or HTN for Hilton), which was noted on their paperwork but which was not used anywhere on the vehicle.

Broadly speaking, the 101s were delivered to the MoD at a rate of between 700 and 800 a year. They were divided between the Ashchurch and Hilton depots, although not always evenly. In addition, twenty-four vehicles were delivered to Land Rover

58 AM 84 was one of the LHD vehicles delivered to the RAF. LAND ROVER

agents in Germany between 1976 and 1978, en route to the Berlin Brigade. Vehicles destined for the RAF also passed through Ashchurch or Hilton, but there is no record of whether those for the Royal Navy did so or not.

There are several unresolved issues about deliveries of the 101s. For example, military vehicle historian Geoff Fletcher points out that some of the dates in surviving records do not seem to match up. He cites the example of 75 GJ 17 (964-00448A), a late LHD 24V vehicle that left the despatch department at Solihull on 16 March 1978 and was recorded as delivered to Ashchurch. However, the army records show that it was receipted into Ashchurch on 13 May 1980 – more than two years later. If the recorded dates are correct, it must have been diverted somewhere and that somewhere was not recorded. Perhaps it went to BAC at Stevenage and was intended for conversion to a Rapier tractor, but for some reason remained unconverted. If so, the entry in the Solihull records should have been amended to show the change of destination, but for some reason was not.

Errors in the records are also a problem. No build information for 956-00983A was recorded in the despatch department records at Rover, which would suggest that it was not built. However, an error in military records strongly suggests that this chassis was built and became 74 FL 26. (The 'error' is that both 73 FL 99 and 74 FL 26 are recorded as having chassis number 956-00958A. 73 FL 99 was probably correct,

HOW MANY IN BRITISH SERVICE?

These figures include pre-production models numbered within the production sequences, but they do not include the prototype vehicles. Some pre-production vehicles were transferred between Land Rover and MVEE, and it is not yet possible to be certain about the end-users of some export vehicles. These difficulties complicate the position, so these figures should be considered as good approximations only.

End-users of 101 types in British service

Type	Built	Not to UK MoD	Total not to UK MoD	UK MoD Total
956	1,321	Rover (5); Leykor, South Africa (1); ROF Nottingham (1); Unknown (2)	9	1,313
959	581	Dubai (13); cancelled Dubai orders (1); Oman (42); UAE (25); Egypt (3)*; Unknown (3)	87	494
961	170	Rover (3)**	3	167
964	473	Rover (3); Unknown (3); Luxembourg (61)	67	406
	2,545		167	**2,380**

* The Egyptian vehicles are assumed to be 959-series, but it is possible that they were actually 964-series' chassis. The 'unknown' three may also have gone to Egypt.

** Note that two 961-series chassis used for trials in Australia were returned to British service.

The vehicles believed to have been retained by Rover and counted in these totals were:

956-00002A, 956-00003A, 956-00016A, 956-00167A and 956-00168A
961-00001A, 961-00012A and 961-00019A
964-00001A, 964-00002A and 964-00003A

The vehicles delivered overseas are listed in Chapter 6.

and 74 FL 26/ 956-00983A would follow neatly from 74 FL 25/ 956-00982A. The error may have been Rover's: perhaps both vehicles had a chassis plate stamped with the same number, which would explain why 956-00983A was never recorded as passing through the despatch department!)

INITIAL UNIT ALLOCATIONS

Military vehicle historian Geoff Fletcher has researched the initial allocations of the 101s, and has come up with the data provided here. Note, of course, that these were initial allocations only. There were many later changes as Britain's armed forces evolved and restructured, and the only way of tracing the life of an individual vehicle is from the record cards which, for army vehicles at least, were at the time of writing held at the Royal Logistics Corps Museum in Deepcut, Surrey.

Infantry Battalions

The 956-series 101 One-Tonnes were used to equip the Support Company in these Battalions, and typically carried the MILAN anti-tank guided missile and the 81mm mortar.

- The twenty-six Regular Infantry Battalions based in the UK each had eight GS vehicles; seven had no winch and one was equipped with a winch.
- The twenty Territorial Army Infantry Battalions with NATO roles each received ten GS vehicles.

Royal Artillery

There were three Royal Artillery Light Regiments in the UK: one supporting 6 Field Force, one supporting 7 Field Force and the third supporting the Commando Brigade. These were all supplied with 956-series vehicles.

Each regiment received eighteen winch-equipped vehicles for use as light gun tractors, plus eighteen vehicles without a winch for use as limber vehicles. Each gun battery consisted of six guns and, in theory, each gun had one gun tractor and one limber vehicle.

In addition, 961-series 24V vehicles went to Royal Artillery Regiments for two different roles. In theory, each battery received two vehicles as command posts to house the field artillery computer equipment (FACE). In practice, only one UK artillery regiment was ever fully equipped with six vehicles, and this was 47 Regiment at Lille Barracks in Colchester.

Three Royal Armoured Corps Regiments received five vehicles each. These were 3 RTR at Bhurtpore Barracks in Tidworth, the Life Guards at Combermere Barracks in Windsor and 9/12 Lancers at Assaye Barracks in Tidworth.

The 961-series vehicles were also used as quartermasters vehicles at both regimental and squadron level in Armoured and Recce Regiments of the Royal Armoured Corps.

BAOR

In early 1978, thirteen of the BAOR Infantry Battalions were equipped with four 959-series LHD GS vehicles each. Two more battalions were equipped in the same way as the UK-based wheeled battalions. So twelve 959-series vehicles went to the Nuclear Convoy Escort Battalion (1 Battalion, The Royal Scots) and eight went to the 5 Field Force Infantry Battalion. In addition, eighteen winch-equipped 959-series vehicles and eighteen 959-series vehicles without winches were issued to 5 Field Force Light Gun Field Regiment (7 Regiment, The Royal Horse Artillery).

Rapier Regiments

Both RHD 961-series and LHD 964-series vehicles were primarily intended to equip Rapier Regiments that were training in 1977, and each of those regiments received, again in theory, thirty-six vehicles.

The 961-series vehicles went to the sole UK Rapier Regiment (initially 16 LAD Regiment) that was based in Lincolnshire at Rapier Barracks, Kirton-in-Lindsey. The 964-series vehicles went to 12 LAD Regiment and to 22 LAD Regiment. 12 LAD Regiment was based at Napier Barracks in Dortmund and 22 LAD Regiment also went to Dortmund after training in the UK with its LHD vehicles.

The RAF Regiment was also equipped with 24V Rapier versions of the One-Tonne. Each of its LLAD (Low Level Air Defence) Squadrons was equipped in the same way as a Royal Artillery Rapier Battery.

Both the army and the RAF also had a number of Rapier Test Vehicles (there is more about these in Chapter 6).

ABOVE: **The 101 and 105mm light gun were both capable of being lifted by the Puma helicopter as separate loads.** GEOFF FLETCHER

LEFT: **Packed on a medium stressed platform (MSP) in preparation for loading into a transport aircraft, this 101 is accompanied by the gun it tows.** GEOFF FLETCHER

Another 101 and gun on an MSP have made a successful landing after being dropped by parachute, which can be seen in the background. GEOFF FLETCHER

BERLIN BRIGADE 101S

Between 1945 and 1990, the city of Berlin was part of the Federal Republic of Germany (West Germany), even though it was located within the Soviet-occupied German Democratic Republic (East Germany). An autobahn provided the only road link between Berlin and the West.

During the mid-1970s, the occupying Soviet forces had banned the Western Allies from driving military convoys along the Berlin Autobahn, which made it very difficult for the UK to supply its Berlin Brigade (which was funded by the West German Government) with new vehicles. When the Berlin Brigade wanted some 101 Forward Controls, a way round the ban had to be found. The solution was to ship the vehicles out to the British Leyland main dealer in West Germany (Brüggemann in Düsseldorf) and for the dealer to deliver them to Berlin. As the vehicles were not strictly in military ownership until they reached Berlin, this arrangement honoured the letter of the requirement, if not the spirit.

So a total of twenty-four vehicles, in three batches, were delivered to Land Rover agents in Germany between 1976 and 1978. All were LHD 24V types. They were ordered from Land Rover by RAOC (Germany), probably via the Berlin Senät who paid under war reparations. The first batch of fifteen vehicles had order nos 35000 to 35014; the second batch of three vehicles had order nos 35015 to 35017A; and the third batch of six vehicles had order no. RWP 100 RAOC Germany. All were ordered with standard bronze green paint, charcoal trim, drab full-length canvas hoods, fume curtains, safety harnesses and hand-throttle controls. At least one is believed to have been repainted, or partially repainted, in the blocky grey, brown and white 'urban camouflage' characteristic of Berlin Brigade vehicles, but the rest probably served in the usual NATO green.

The first fifteen went during 1976 to Brüggemann, followed by three more to the same agent in 1977. The final six were delivered during 1978 to BL Germany (overseas sales agencies had by this time been re-organized).

The three batches were as follows:

964-00284A to 964-00298A: registered 58 XB 20 to 58 XB 34 in arbitrary order
964-00380A to 964-00382A: registered 58 XB 38 to 58 XB 40 in chassis number order
964-00466A to 964-00471A: registered 27 XH 81 to 27 XH 86 in chassis number order

SERVICE LIFE

The 101s spent many years in readiness for operational activity, and many of them saw use in the endless round of exercises designed to keep NATO forces in West Germany capable of responding rapidly to the threat of invasion by Soviet forces across the border with East Germany. This activity obviously declined after the thaw in East–West relations that began in 1990. Large numbers saw service with Royal Marines Commando units.

A number were also allocated to UNFICYP (United Nations Peacekeeping Force in Cyprus), where their everyday activities were mainly associated with the supervision of ceasefire lines and the maintenance of a buffer zone. They were also used for various humanitarian activities in the area. One group,

74 FL 46 (956-00928A) was pictured during wading trials at ATTURM, near Instow in Devon. It carries the markings of 29 Commando Regiment. ATTURM

80 ■ THE 101 IN BRITISH MILITARY SERVICE

A 101 gun tractor stands by as the crew of a 105mm light gun deploy the weapon during a demonstration. LAND ROVER

configured as Rapier tractors and operated by 6 Wing RAF, was used for the defence of USAF airfields in the UK after 1983.

Several 101s were deployed to the Falkland Islands in 1982, when British troops were sent to retake the islands that had been occupied by Argentinian forces. An even more high-profile deployment came with Operation GRANBY in 1991, when British troops and vehicles were sent to Kuwait in support of the legitimate government of a country that had been invaded by the Iraqi dictator Saddam Hussein. A few also went to the former Yugoslavia in 1995 with the British contingent attached to UNPROFOR (United Nations Protection Force). These belonged to 24 Airmobile Field Ambulance, and several ambulances remained in theatre when UNPROFOR transformed into IFOR, the peace Implementation Force and then the smaller SFOR, the Stabilization Force.

74 FL 05 (956-00964A) was pictured waiting its turn during a military demonstration at Rushmoor, near Aldershot. The door marking indicates that this vehicle belonged to the lead battery. DUNSFOLD COLLECTION

THE 101 IN BRITISH MILITARY SERVICE — 81

BELOW: **Painted in United Nations white, 69 FL 37 awaits deployment overseas with a large convoy of other military vehicles.** GEOFF FLETCHER

BELOW: **A 101 ambulance, 72 GJ 00, is pictured in UN livery amid the ruins of a town during the Bosnian deployment. It belonged to 24 Field Ambulance.** LAND ROVER

By this stage, however, the numbers of 101s in service had dwindled. Although a few of the 101s had been withdrawn during the 1980s, often after sustaining major damage (and small numbers were withdrawn in the 1970s), the big surge in withdrawals began in the early 1990s. By that stage, the earliest examples, delivered in 1975, had been in service for 15 years, and military policy meant that was time for them to be replaced.

Nevertheless, replacement was not quite as straightforward as the UK MoD might have liked. The search for a replacement fleet for the One-Tonnes had begun in the mid-1980s, and Land Rover itself had submitted a new forward-control vehicle, internally codenamed Llama, for the trials. The major drawback of this vehicle was that it was not available with a diesel engine (although a new one was under development), and it lost the contract to a modified Dodge 50-series truck that was built in the UK with a Perkins diesel engine as the Reynolds Boughton RB44.

The contract for 846 RB44s was placed in June 1988, but early examples delivered suffered from braking problems. The MoD continued to monitor the position, and in 1992 suspended deliveries. In 1994–95, RB44s were put through a major re-work programme to solve the braking problems, and meanwhile the MoD had embarked on an

82 ■ THE 101 IN BRITISH MILITARY SERVICE

unscheduled refurbishment programme for some of its remaining 101s, in order to keep an adequate vehicle fleet in use. Some of the refurbishment work was done in an area of the corporation bus garage in Portsmouth that had been specifically allocated for the task. Many of the last 101s remaining in service were allocated to Royal Marines Commando and Parachute Regiment units, which still had a

ABOVE: **Land Rover hoped to win the 101 replacement contract with its specially-developed Llama forward control. This was one of the military trials vehicles.**
LAND ROVER

BELOW: **A 101 stands next to the vehicle that replaced it in many units, but was never as well liked – a Reynolds-Boughton RB44.**
GEOFF FLETCHER

requirement for this type of vehicle and found the RB44s unsuitable.

A generous handful of 101s remained in UK military service in the mid-1990s, but the final vehicles seem to have been withdrawn in 1998 – 20 years after the last examples were delivered.

Large numbers of British military 101s were sold off on the civilian market in the 1990s, but some vehicles never made it into civilian life, being either written off or cannibalized for spares while in service. A fairly typical example of the latter was 72 FL 39 (964-00317A), which was struck off in August 1994 by 237 Signal Squadron.

THE QUESTION OF RELIABILITY

There is no doubt that most military users of the 101 thought it was an excellent vehicle, although there is equally no doubt that there were some who had particular gripes. The uncomfortable driving position was a major one of those. But individual opinions were of little use to the British Army.

What it wanted was an objective report on the 101's in-service reliability. So Vehicles Branch, REME was given the job of finding out how good or otherwise the 101 really was. The top priority was to discover whether the problems identified with the pre-production vehicles trialled a few years earlier had been satisfactorily rectified. In practice, the report was not published until April 1978, by which time the final vehicles were about to be delivered, but if they had a normal service life of 15 years, the report's findings would be valid for some time to come.

The initial data were collected between 1 August 1976 and 31 July 1977, but that 12-month period was subsequently extended by a further 5 months to the end of 1977. There were 180 vehicles in the survey, based at nine different UK and BAOR locations, and together they were assessed over a distance of more than 970,000 miles (1,560,000km), which averaged out at about 5,390 miles (8,672km) for each vehicle. Most of that was road mileage: rather less than 9 per cent was on tracks or across country.

Back in May 1968, the reliability requirement for the 101 had set 7,000 miles (11,265km) as an acceptable 'mean distance between minor failures', and 60,000 miles (96,551km) as the 'mean life of major components excluding the engine'. The mean figure for engine life was set at 30,000 miles (48,280km). All these figures would have seemed low to the buyer of a civilian Land Rover at the time, but of course military users do subject their vehicles to more than their fair share of hard usage.

Another important figure that REME used for its assessment was the MBDF or mean distance between failures. On the pre-production vehicles, this had been just over 1,000 miles (1,609km), and REME had predicted that 1,560 miles (2,510km) should be attainable if production vehicles rectified twenty-five

63 FL 04 (956-00331A) and 63 FL 41 (956-00368A) were pictured during a military demonstration at the Rushmoor arena. The vehicles belonged to 7 RHA at the time.
DUNSFOLD COLLECTION

recurrent failures and a further ten quality control issues were addressed. That figure proved a little optimistic; the vehicles in the survey actually achieved an MDBF of 1,456 miles (2,343km).

There was another disappointing result in the report, and that was the battle-worthiness failure figure. Translated, this means a failure that prevents the vehicle from carrying out its battlefield role, independently of whether or not it is still roadworthy. At 5,880 miles (9,463km), this made clear that REME's mechanics would have plenty of work to do on the 101s. The roadworthiness figure, where roadworthiness means legal requirements for continued operation, was a miserable 1,282 miles (2,063km). Illustrative of the problems was that three engines had to be replaced during the data-collection period. One had seized when an oil-cooler pipe fractured, and two failed through excessive camshaft wear. The latter was not uncommon, either: five more reports of the problem had also been received outside the confines of the study itself.

Particularly damning was the report's assessment that any given example of the 101 had only a 95 per cent probability of completing a battlefield mission of 75 miles (121km). Six areas set out in the condition of acceptance were unsatisfactory: these were failures of the throttle linkage, the throttle cable, the gearbox and/or transfer box selector, the hub oil seals, the universal joints on the propshafts and the steering ball joints.

A second section of the report listed what it called 'new trends occurring' in production vehicles; these were problems that had not arisen during the pre-production trials but were apparent from the in-service study of those 180 vehicles. This did not paint a pretty picture, either. Problems were reported in the electrical system (ninety-seven cases), footbrake (sixty-seven), body and tilt (sixty-three), transmission brake (fifty), ignition system (twenty-three), clutch (twenty-one), transfer box (sixteen), oil cooler (eleven), carburettors (ten), fuel tank (nine), steering box (eight) and differential lock (seven). In addition, there were problems with an imbalance in the braking system when the vehicle was unladen.

The REME report concluded that the 101 was not going to meet its expected reliability criteria. 'Despite changes in hand,' it read, 'it is unlikely that the target mean distance between minor failures of 7,000 miles (11,265km) will ever be achieved.'

Nevertheless, a number of minor modifications were made to the vehicles and to the assembly process at Solihull to address these issues, and the 101 is not remembered as a particularly unreliable vehicle. In fact, most users with direct experience of it tend to remember it as very reliable and rugged.

VEHICLE IDENTIFICATION PLATES

At first sight, there appears to be no logic to the relationship between the chassis numbers of the 101s and their military serial numbers. Broadly speaking, that impression is correct, although the system was certainly not random.

69 FL 66 (956-00882A) was pictured at Larkhill Artillery Day in July 1980. Although marked for 59 (Asten) Training Battery of 17 Training Regiment at Woolwich, it was in fact allocated to 56 Squadron RCT at Woolwich. GEOFF FLETCHER

The military reserved large batches of serial numbers in advance of deliveries, and vehicles were allocated a military registration before they were built, the plates being applied at the Rover factory in Solihull. This sounds simple enough, but a number of factors affected what happened in practice.

The first problem was that the serial numbers were allocated when the contracts were issued. The contracts were amended before the vehicles were delivered, and the batches of numbers allocated consequently turned out to be either too large or too small. So once a group of registration numbers had been used up, more had to be found, and they would most likely be many hundreds of numbers further on, after a group originally reserved for a different type of vehicle.

Every production 101 left Solihull with a metal plate attached by pop-rivets to the radiator ducting (which runs between the seats in the cab) on the passenger's side. This plate was stamped with the chassis number, the recommended maximum towing capacity, and the unladen weight and maximum front and rear axle weights, all in both kilogrammes and pounds. This was the manufacturer's plate, and broadly conformed to the type used on Land Rovers for the civilian market.

A second or military identification plate was added next to the manufacturer's plate, this time attached by slot-head screws. It appears that these second plates were added at Solihull before delivery to the military. They were stamped with the chassis number, the contract number, the military serial number and the military asset code number. Further spaces were available for the CES number (the initials stand for Continuous Engineering Support) and for records of major repairs carried out during service.

MILITARY SERIAL NUMBERS

As the system used for vehicle serial numbers changed during the service life of the 101 One Tonne, a few words of explanation will prevent confusion.

Before 1982, the army, Royal Navy and RAF all had separate serial number systems. From 1982, a new 'tri-service' series was used, and this was allocated to all vehicles as the contracts were placed.

The pre-1982 Royal Navy registrations used a six-digit number with RN as the centre pair, and the RAF used a similar system with A as the first letter of the centre pair. The numbers always started at 0001, and were split around the central digits. So 00 RN 01 would have been the first number in the Royal Navy series, 00 RN 02 would have been the next and so on up to 99 RN 99. This, at least, was the theory: in practice, many RN and a small number of RAF numbers were reused over time.

The army used the same six-digit system, but as the largest user of vehicles it had more complicated arrangements. To simplify the issues, the lists below are confined to the numbers found on 101s. Note that it was not uncommon for a serial number to be changed if a vehicle was re-allocated.

BT, e.g. 54 BT 07. The BT series was used for vehicles not acquired under contract, and included transfers from MVEE and from the Royal Navy or RAF.

FL, e.g. 68 FL 39. The FL series was for vehicles ordered under contracts issued in the 1971–72 financial year. Most British Army 101s were numbered

This was the standard layout of identification plates on a 101, in this case a LHD 964-series vehicle. AUTHOR

PAINT SCHEMES FOR THE UK 101S

All the 101s for the UK MoD were painted at the Rover factory in gloss deep bronze green, the standard peacetime colour for non-operational vehicles. However, most probably lost this soon after allocation to units, when it would have been overpainted in whatever scheme was appropriate for the local conditions.

Vehicles operated by the UK armed forces were often overpainted with NATO matt green IRR (infra-red reflective) paint, which made them less visible to infra-red night vision equipment. This was available as a spray paint with reference number COSA no. H1/8010-99-224-8906 or as a brush paint with reference number COSA no. H1/8010-99-224-8907. The only permissible additive was thinner paint with reference number COSA no. H1/8010-99-942-7564; other thinners were liable to reduce the IRR properties of the paint. It appears that the ambulances were mainly delivered with IRR paint already applied.

Many vehicles had a camouflage scheme of black and green patches. Those that served with the peace-keeping force in Cyprus were painted all-over white, and an all-over white scheme was used on some vehicles involved in joint exercises in Scandinavia and the Arctic. White was used again for vehicles attached to UNPROFOR (United Nations Protection Force) in Bosnia-Herzegovina in the mid-1990s. Many vehicles deployed on Operation GRANBY in the Gulf were overpainted in an all-over sand colour or in a mixture of sand and pink patches, sometimes with darker sand or grey camouflage patches.

One of the Berlin Brigade vehicles is thought to have been repainted in the special urban camouflage used in that city.

ABOVE: **69 FL 29 (961-00148A) shows an unusual in-service camouflage pattern, although it was actually photographed in the UK. The vehicle was being used as a recce unit for the Ace Mobile Force at the time.** GEOFF FLETCHER

LEFT: **Wearing fairly typical 'home' camouflage is this RAF 101.** GEOFF FLETCHER

in this series, even though deliveries continued right through to 1978.

GJ, e.g. 73 GJ 32. The GJ series was used for vehicles ordered in 1975–76. In practice, most of the ambulances were built on chassis ordered in this period, even though they did not enter service until the 1980s.

KB, e.g. 42 KB 59. The KB series was issued for vehicles that were ordered between August 1982 and November 1983.

Serial Numbers of the British Military 101s

These figures include the pre-production vehicles (i.e. those numbered in the production chassis sequences) but do not include the prototypes. The information in these tables combines research by Geoff Fletcher of the Military Vehicle Trust and by Les Adams of the 101 Forward Control Club and Register. Problems in some of the surviving records mean that the tables represent the most accurate list that can be achieved at present.

956-SERIES VEHICLES

(Army Numbers)

03 SP 67	1
03 SP 68	1
03 SP 70	1
60 FL 32 to 67 FL 49	718
69 FL 40 to 69 FL 82	43
70 FL 83 to 70 FL 85	3
72 FL 79 to 74 FL 10	132
74 FL 13 to 74 FL 73	61
76 FL 64 to 76 FL 66	3
SUB-TOTAL	963

(Royal Navy Numbers)*

73 RN 00	1
SUB-TOTAL	1

(Royal Air Force Numbers)

58 AM 47 to 58 AM 48	2
81 AM 35 to 81 AM 43	9
81 AM 81 to 81 AM 82	2
00 AM 72 to 01 AM 06	35
01 AM 07 to 01 AM 41	35
SUB-TOTAL	83

(Tri-Service Numbers)

71 GJ 51 to 73 GJ 57	204
(73 GJ 00, 53 and 54 cancelled)	
75 GJ 50 to 76 GJ 02	53
18 KJ 35	1
18 KJ 43	1
34 KJ 79	1
34 KJ 95 to 34 KJ 96	2
48 KJ 49 to 48 KJ 50	2
55 KJ 69 to 55 KJ 70	2
SUB-TOTAL	266
TOTAL	**1313**

* (Note that the Royal Navy also had 02 RN 37, which was an unidentified pre-production vehicle and probably carried another registration number at some time.)

959-SERIES VEHICLES

(Army Numbers)

03 SP 72/77 FL 28	1
67 FL 50 to 68 FL 38	89
69 FL 83 to 70 FL 72	90
71 FL 40 to 71 FL 67	28
75 FL 31 to 75 FL 38	8
76 FL 51 to 76 FL 63	13
76 FL 67 to 77 FL 06	40
SUB-TOTAL	269

(continued overleaf)

Serial Numbers of the British Military 101s (continued)

(Royal Navy Numbers)

None	SUB-TOTAL	0

(Royal Air Force numbers)

58 AM 49 to 58 AM 54	6
81 AM 31 to 81 AM 33	3
81 AM 44 to 81 AM 76	33
81 AM 83 to 81 AM 84	2
82 AM 33 to 82 AM 50	18
SUB-TOTAL	62

(Tri-Service Numbers)

73 GJ 65 – 73 GJ 99	35
74 GJ 00 – 74 GJ 99	100
75 GJ 00 – 75 GJ 14	15
75 GJ 1	1
75 GJ 18	1
31 KB 97 – 31 KB 98	2
42 KB 59 – 42 KB 68	10
SUB-TOTAL	164
TOTAL	**495**

961-SERIES VEHICLES
(Army Numbers)

68 FL 39 to 68 FL 43	5
68 FL 45 to 69 FL 39	95
74 FL 11 to 74 FL 12	2
74 FL 81 to 75 FL 30	50
03 SP 78	1
05 SP 10 to 05 SP 11	2
Numbers not traced	2
SUB-TOTAL	157

(Royal Navy Numbers)

None	SUB-TOTAL	0

(Royal Air Force Numbers)

58 AM 38 to 58 AM 46	9
81 AM 77 to 81 AM 80	4
SUB-TOTAL	13

(Tri-Service Numbers)

None	SUB-TOTAL	0
	TOTAL	**170**

964-SERIES VEHICLES
(Army Numbers)

68 FL 44	1
70 FL 73 to 70 FL 82	10
70 FL 86 to 71 FL 39	54
71 FL 68 to 71 FL 88	21
71 FL 90 to 72 FL 78	89
75 FL 39 to 76 FL 50	111
03 SP 69	1
03 SP 71	1
04 SP 08	1
58 XB 20 to 58 XB 34	15
58 XB 38 to 58 XB 40	3
27 XH 81 to 27 XH 86	6
SUB-TOTAL	313

(Royal Navy Numbers)

None	SUB-TOTAL	0

(Royal Air Force Numbers)

58 AM 55 to 58 AM 87	33
81 AM 34	1
83 AM 76 to 83 AM 80	5
SUB-TOTAL	39

(Tri-Service Numbers)

75 GJ 17 to 75 GJ 34	18
SUB-TOTAL	18
TOTAL	**370**

KJ, e.g. 18 KJ 35. The KJ series 'tri-service' numbers were theoretically issued for vehicles ordered between September 1989 and July 1990. They were used on some late-issue 101s that had been in store but were needed by Parachute Brigade units that were running short of serviceable examples.
SP, e.g. 04 SP 67. The SP series were used for special projects, and were associated with prototypes under evaluation by MVEE. SP numbers were allocated to the 101 initial trials vehicles.

RAF AND ROYAL NAVY 101S

The majority of British 101s served with the army, although it was increasingly difficult to tell which service a vehicle belonged to after the introduction of the unified serial numbering in 1982 that was used by all three services. In fact, vehicles with earlier registration serials sometimes ended up with a different service: 7 RHA, for example, received a number of 101s with RAF serial numbers.

RAF Vehicles

The RAF's first 101s were probably Rapier tractors for airfield defence duties. From 1978 it also took on a quantity of 956-series (12V RHD) vehicles as transfers from the army, and after 1980 also took some ambulance-bodied variants. Some of these were standard four-stretcher types and others were specialized airfield crash-rescue variants. A small number of signals bodied vehicles also served with the RAF.

In addition, the RAF operated 101 Rapier tractors in defence of USAF airfields in Britain from 1983.

MILITARY ASSET CODES

Ambulances
1053-2750: RHD tropicalized
1053-6750: LHD tropicalized
1054-0750: RHD (FV 19009)
1054-5750: LHD (FV 19010)

FACE Carriers
1851-0750: RHD, 24V

GS 12V
1825-0750: RHD, with winch
1825-5750: LHD, with winch
1840-0750: RHD
1840-1750: RHD, with trailer drive
1840-5750: LHD
1841-0750: RHD, winterized

Intercept Complex
1852-4100: 24V

FFR 24V
1830-0750: RHD, with winch
1830-5750: LHD, with winch
1832-0750: RHD, with winch
1832-5750: LHD, with winch
1850-0750: RHD
1850-5750: LHD

Signals 24V
1831-0750: RHD, with winch
1834-0750: RHD
1834-5750: LHD

Rapier Vehicles
1826-0750: RHD, 12V, with winch, radar tractor for Rapier
1833-0750: RHD, 24V, with winch, launcher tractor for Rapier
1833-5750: LHD, 24V, with winch, launcher tractor for Rapier
3646-0750: RHD Rapier automatic test equipment, field standard 1B
3646-1750: RHD Rapier automatic test equipment, field standard 1A
3646-5750: LHD Rapier automatic test equipment, field standard 1B
3664-9300: LHD Vampire

A total of forty 956-series (RHD, 12V) vehicles were transferred from the army to the RAF in July–August 1978. Two were similarly transferred in July 1977 and one more in June 1979, making a grand total of forty-three. None of these vehicles entered service with its original army registration number except the last two, 72 FL 79 and 73 FL 50.

Royal Navy Vehicles

Large numbers of 101s were used by Royal Marines Regiments, but these always carried army or unified registration serials. The Royal Navy appears to have registered only two 101s for its own use. These were 73 RN 00 (a 956-series GS truck) and 02 RN 37 (an otherwise unidentified vehicle that appears to have been on exercise in Norway with the Royal Marines of 3 Commando Brigade).

RAF 101s

Original	RAF	Date	Original	RAF	Date
61 FL 37	94 AV 43	July 77	**67 FL 46**	36 AJ 85	Aug 78
61 FL 39	94 AV 44	July 77	**67 FL 48**	37 AJ 04	Aug 78
62 FL 77	35 AJ 62	Aug 78	**67 FL 49**	36 AJ 97	July 78
66 FL 83	75 AJ 65	Aug 78	**69 FL 41**	36 AJ 98	July 78
66 FL 93	35 AJ 60	Aug 78	**69 FL 42**	37 AJ 00	July 78
67 FL 01	36 AJ 77	July 78	**69 FL 43**	36 AJ 87	July 78
67 FL 02	36 AJ 89	July 78	**69 FL 44**	37 AJ 02	July 78
67 FL 11	36 AJ 81	July 78	**69 FL 45**	36 AJ 73	July 78
67 FL 13	36 AJ 82	July 78	**69 FL 56**	36 AJ 94	July 78
67 FL 14	35 AJ 59	Aug 78	**69 FL 59**	36 AJ 84	July 78
67 FL 17	35 AJ 53	Aug 78	**69 FL 60**	35 AJ 66	July 78
67 FL 26	36 AJ 78	Aug 78	**69 FL 65**	36 AJ 90	July 78
67 FL 29	35 AJ 61	Aug 78	**69 FL 69**	36 AJ 92	July 78
67 FL 30	36 AJ 86	Aug 78	**69 FL 73**	35 AJ 56	July 78
67 FL 31	36 AJ 75	July 78	**69 FL 76**	35 AJ 68	July 78
67 FL 32	36 AJ 74	July 78	**69 FL 77**	30 AJ 96	July 78
67 FL 33	35 AJ 58	Aug 78	**69 FL 78**	35 AJ 52	July 78
67 FL 34	37 AJ 05	July 78	**69 FL 79**	37 AJ 03	July 78
67 FL 38	35 AJ 55	Aug 78	**69 FL 81**	36 AJ 88	July 78
67 FL 42	36 AJ 99	July 78	**72 FL 79**	94 AM 48	Aug 78
67 FL 43	35 AJ 69	Aug 78	**73 FL 50**	49 AS 66	

CHAPTER SIX

BRITISH 101 VARIANTS

From the very beginning, the MoD had envisaged a number of roles for the One-Tonne Forward Control. As Chapter 1 explains, FVRDE made a number of scale models as early as 1967 to demonstrate variants that included a personnel carrier, WOMBAT portee, four-stretcher ambulance, missile launcher and mortar carrier.

The vast majority of 101s delivered to the British armed forces were configured as GS trucks. However, more than 500 were delivered as chassis-cabs,

> ## VISIONS OF THE FUTURE
>
> By the time Rover was in a position to offer the 101 for sale to customers other than the UK MoD, there had already been trials of the vehicle in a number of configurations. So the first Rover sales brochure (issued in November 1974) and its second edition showed 'artist's impressions' of a number of these. They had been drawn up by a local commercial artist called Pittaway.
>
> The Pittaway drawings showed the vehicle kitted out not only for its role as a gun tractor but also as an airfield crash tender, as a WOMBAT portee, a Beeswing anti-tank missile launcher, a carrier of electronic equipment (the illustration may have been intended to represent a fire control system such as FACE), a tractor for the Rapier surface-to-air missile system and an ambulance. Neither the airfield crash tender nor the Beeswing portee entered British military service, either, although the Beeswing was supplied to some overseas military forces, as Chapter 7 explains.
>
> **One that got away: the 101 never did see service as an airfield crash tender.** LAND ROVER

> ### WINCH OR NO WINCH?
>
> Figures calculated by Geoff Fletcher suggest the following figures for UK MoD deliveries of GS trucks. Totals include pre-production but not prototype vehicles.
>
> | Chassis series | With winch | Without winch |
> |---|---|---|
> | 956 | 279 | 775 |
> | 959 | 221 | 100 |
> | 961 | 63 | 104 |
> | 964 | 180 | 190 |

and the intention was that these should be bodied when they were required. These vehicles were put into storage at Ashchurch military vehicle depot along with a number of other unissued vehicles. They began to emerge in 1980, as the new bodying programme began. Many became ambulances; some were bodied as signals vehicles (although not all were used in that role); and a few simply became GS trucks – with some differences from earlier production versions.

From the time they began to enter service in 1975, the general service trucks were used for the wide variety of duties associated with all GS vehicles. Many were modified by user units to meet particular requirements. All this produced a fascinating variety of One-Tonne types, and the purpose of this chapter is to shed some light on what they were.

AMBULANCE

An ambulance was among the variants of its new Forward Control vehicle that FVRDE envisaged in 1967, and a scale model was built to show the anticipated four-stretcher configuration. As early as August 1969, prototype 101/FC/2 was used to carry a wooden mock-up of the field ambulance body. Nothing more seems to have happened for a time, but the 1974 sales' brochure for the 101 One-Tonne included an artist's drawing of a 101 ambulance, which had a back body that shared some features with that of the 2/4-stretcher ambulance body then being built by Marshall's for the normal-control Series III 109. This may have been a purely a speculative idea of how an ambulance body might look.

Nevertheless, by 1975 the MoD was ready to go ahead with an order for 101 ambulances, although there was clearly no hurry for these to be delivered. The first physical manifestation of the renewed interest came when MVEE constructed a mock-up ambulance body on 956-00005A (03 SP 70). Meanwhile, the requirement for a new battlefield ambulance was issued as GSR 3525/1.

The next stage was a contract for their chassis, which was placed with Land Rover in March 1976 as number WV 12074. The vehicles were built as chassis-cabs, and were put into storage after being allocated temporary numbers in the CC series reserved for such vehicles. (They received ordinary military serial numbers after being bodied; for example, 76 CC 33 became 18 KJ 43.) FVRDE proceeded to the prototype stage in 1976, when two standard 12V GS vehicles delivered to the military transport depot at Ashchurch in Gloucestershire in the first half of the year were pulled from storage and sent to Marshall's of Cambridge to be converted.

One had left-hand drive (959-00033A, 67 FL 71) and the other right-hand drive (956-00478A, 64 FL 51). Marshall's had been asked to produce two different versions of the same body, and so they built a basic four-stretcher air-conditioned ambulance onto the LHD chassis while the RHD chassis had a body with NBC (nuclear, biological and chemical) protection – in effect, a pressurized air system in the ambulance body. Both vehicles were completed and evaluated during 1976, but it was December 1979 before Marshall's received the contract, number FVE 21A/156, to build the bodies for a further large batch. There were to be some for the RAF as well as for the army, all with a four-stretcher design. Only tropicalized versions of the battlefield ambulance would have the air-conditioning that had been built into the LHD prototype; others would have a back body without it.

In the mean time, there had been several changes in the military requirement for the ambulances, and the original contract had been amended in detail

64 FL 51 (956-00478A) was the prototype RHD ambulance, equipped with a pressurized body to guard against NBC hazards. It is seen here in a Marshall's publicity picture, still wearing its FVRDE wing number. The twin driving lamps under the front bumper were not a standard fit. MARSHALL'S

several times. The original order for 519 looked very different by the time the first production ambulances were built in 1980, and it would change again before the contract was fulfilled. As already explained in Chapter 5, sixty-nine chassis originally intended as ambulances were not so bodied, and Marshall's were asked to build 'pattern' GS bodies for them instead. A further bizarre complication was that between fifteen and eighteen LHD vehicles, which had already been completed as GS trucks, were withdrawn from storage and were stripped back to chassis-cab condition in May 1980 at 93 Vehicle Depot Workshop, REME, so that they could also be fitted with ambulance bodies. The logic of all this was no doubt clear at the time, but is impossible to reconstruct 35 years later.

The final user trials for the 101 ambulances took place in 1980 after the first few had been completed, and deliveries continued until the last examples entered service during 1984. However, there are still disputes about how many ambulances were built. The best figures available show that 450 'production' examples were constructed of the 519 originally ordered; to these figures should be added the two prototypes.

More certain is that all the 101 ambulances were on 12V chassis, which were modified from standard in only small ways. They had uprated dampers to cope with the extra weight of the ambulance body, and they had higher output alternators and a split-charge electrical system. The cabs had the standard layout with two seats, which were for the driver and the medical attendant, who would travel in the back when there were casualties on board.

The back body of the ambulances was double-skinned and well insulated, and had two windows in each side and one more in each of its twin rear doors. All six windows had black-out blinds. The inner surfaces were all painted pale green (the colour is called Sea Foam), as was already British military practice, because this colour supposedly helps to calm medical cases. The body was laid out to carry four stretchers, but the upper pair of stretchers could be folded back against the body sides so that the lower pair could be used for seated casualties. In theory, three casualties could then be carried on each side of the vehicle, facing inwards. All 101 ambulance bodies were properly ventilated and had their own heating, and the tropicalized variants were also air conditioned. All ambulances had an oxygen supply fitted in the rear.

The ambulances were able to carry up to four stretchers. 72 GJ 41 was pictured at a special display held in 1985 at Silverstone, to celebrate 100 years of the motor car. It is wearing typical disruptive-pattern camouflage. AUTHOR

All except two of the RAF ambulances had left-hand drive, and most were delivered to the RAF Regiment Squadrons at Gutersloh and Laarbruch in Germany. Their special features nevertheless seem to have been tried out on the right-hand drive ambulance prototype 64 FL 51, which was modified for the purpose and was probably evaluated by MVEE. Like all RAF vehicles, they were fitted with an engine pre-heater and a Niphan trickle-charger for the battery; the 240V socket for this was mounted above the left-hand rear lights, and was also fitted to the One-Tonne ambulances used by the Army Air Corps. An ELCB (earth leakage circuit breaker) was fitted as a safety measure. Instead of a heater for the back body, the RAF ambulances had a radio installation, and all carried resuscitation equipment, with stowage for oxygen and nitrous oxide gas cylinders.

Only twenty-two ambulances appear to have carried RAF registration numbers. After 1982, new examples in RAF service carried numbers in the unified tri-service GJ series. Many years later, 82 AM 45 (959-00529A) was donated to Uruguay as a gift by the Defence Attaché in Montevideo, but it is not clear whether it ever entered military service with that country.

All ambulances carried red cross markings on a white background on the sides, rear and roof. These markings could quickly be concealed if it was necessary to reduce the risk of the vehicles being seen. Each red cross could be covered by a hinged panel that was normally stowed against the adjacent body panel; releasing a catch allowed body-colour panels to drop down over the side panels, while the one on the roof had to be swung into place by climbing up onto the top. Record cards suggest that all the vehicles were delivered in IRR (infra-red reflective) paint, although several different colour and camouflage schemes were used in service.

The first major operational deployment of the 101 ambulances was to the Falkland Islands on Operation *Corporate* in 1982. The vehicles that went on this operation were painted in the usual green and black

BRITISH 101 VARIANTS ■ 95

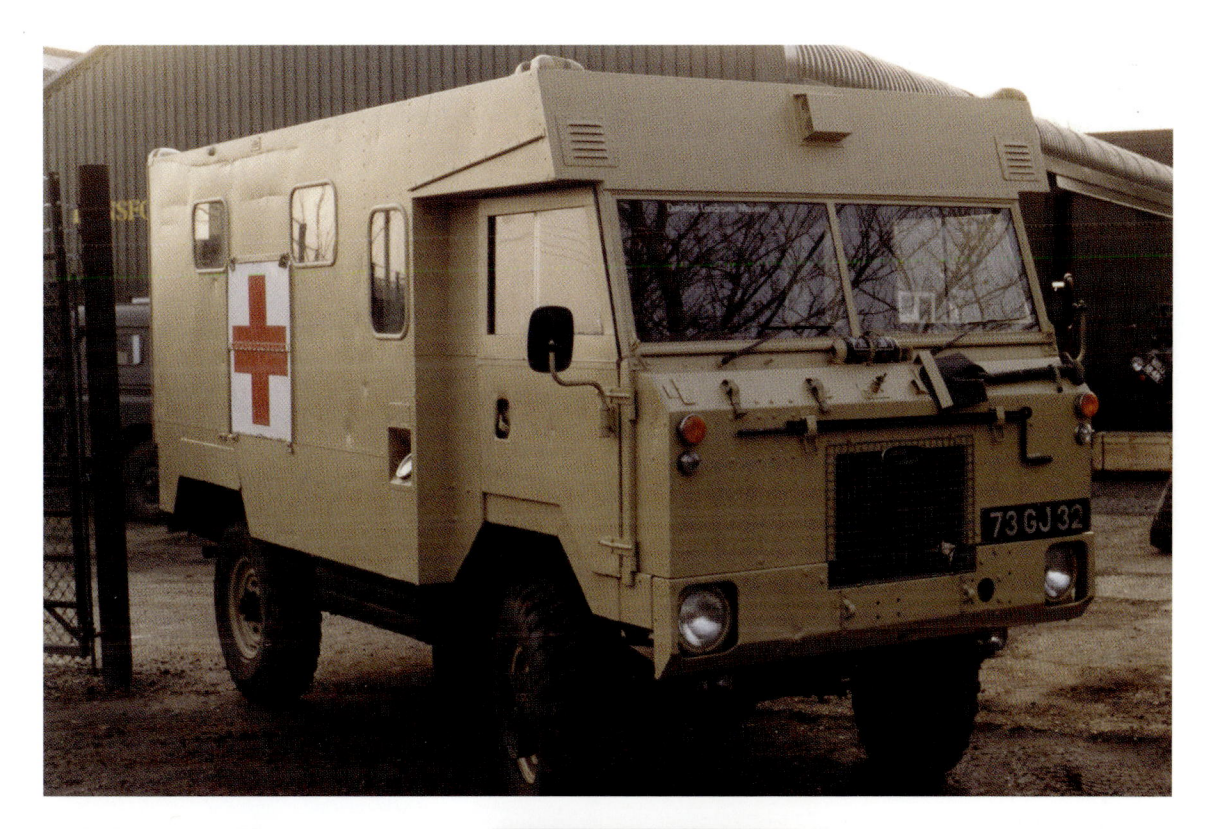

73 GJ 32 (956-01228A) served with 2 Scots Guards Regiment and was one of only two 101 ambulances that came back from the Falkland Islands deployment. It was then carefully rebuilt by Dunsfold Land Rovers (now DLR) before re-sale on to the civilian market. The vehicle is finished in sand colour, which was not used in the Falklands campaign but would be used in the Gulf.
AUTHOR

BRITISH 101 VARIANTS

LEFT: **'Warning – one person only allowed on roof'.** Details like this plate are often missed in a casual look at a 101 ambulance, especially if they have been covered by several layers of paint. AUTHOR

OPPOSITE PAGE: 75 GJ 63 was pictured on duty at Middle Wallop Army Airfield in the mid-1980s. The choice of white livery appears to have been made by the airfield; the vehicle had not served in any overseas theatre where the colour scheme was needed. The fire tender alongside is based on a Reynolds–Boughton RB44 – the same chassis that was bought in quantity some years later as a replacement for the army's fleet of 101s. AUTHOR

This collection of plates on 73 GJ 32 are the military type plate, the 'T' plate indicating that the vehicle has been tropicalized, and the IRR plate indicating that it has at some point been painted with infra-red reflective paint. AUTHOR

camouflage; just two of them returned to the UK, one of them without its engine and gearbox, which had been removed for use in another vehicle. Several were next used on a peace-keeping deployment to Cyprus, and these were painted in the UN colour of white. The ambulances attached to infantry battalions in Cyprus were painted in two-tone sand and green, and so were those which became OPFOR (Opposition Forces) vehicles on the BATUS training ranges in Canada.

Some ambulances went out to Kuwait on Operation GRANBY in 1991 and were painted sand colour; as the repainting was done in something of a hurry, there were several variations, and some vehicles were hand-painted with a combination of sand and dark sand. Many of those that went to Croatia in the mid-1990s with the British peace-keeping contingent were again painted in United Nations white. They also gained temporary seven-digit registration numbers that began with the letters UNPF.

The 101 ambulances remained in service until the late 1990s, when they were replaced by the new Land Rover Defender XD130 ambulances. The final examples were probably withdrawn in 1998.

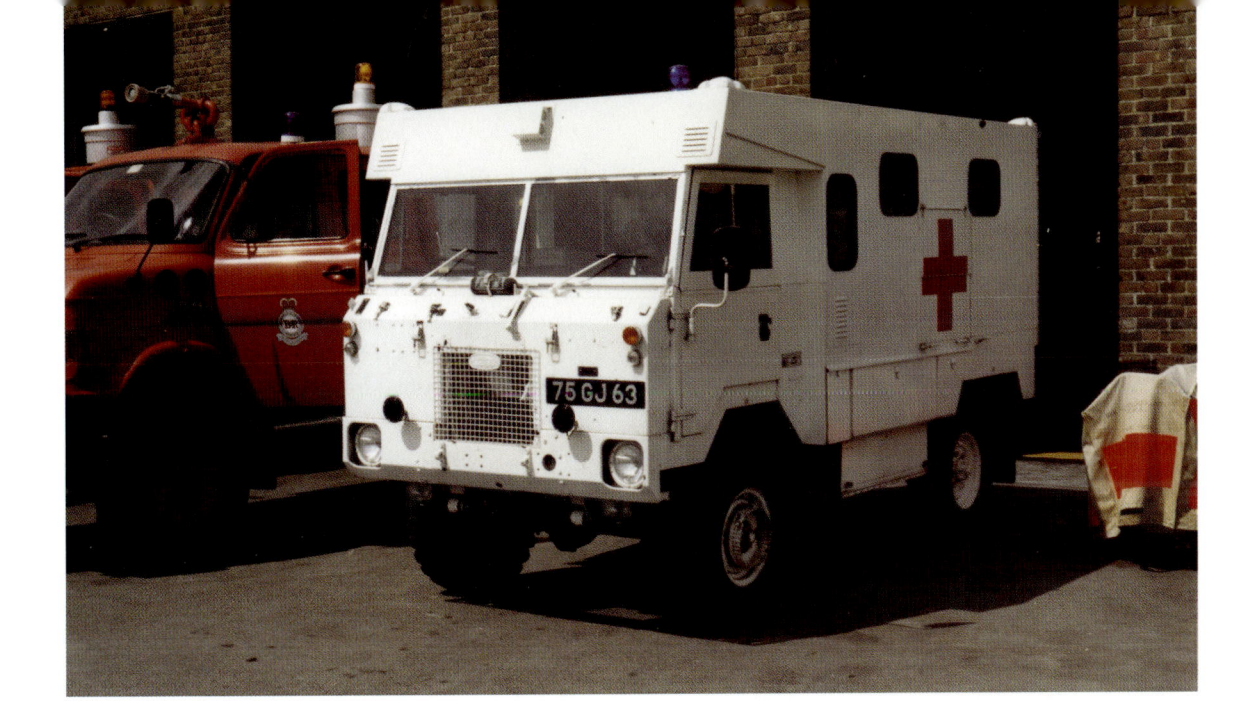

THE MARSHALL'S AMBULANCES

There were 450 'production' ambulances and two prototypes. The 450 production vehicles appear to have consisted of:

256 RHD for the army
151 LHD for the army from chassis-cabs
21 LHD for the army converted from GS vehicles
20 LHD for the RAF
2 RHD for the RAF

Approximately twenty were registered in the FL series, including the two prototypes. The largest number was in the GJ series, of which 407 have been traced. There were two in the KB series, and the twenty-two RAF ambulances were all initially registered in the AM series, although the two RHD vehicles were later given new numbers in the BT series.

Pictured when new by its makers, Marshall's of Cambridge, this LHD ambulance was 73 GJ 81 (959-00543A). The red cross on the roof, not normally visible, is clear here. MARSHALL'S

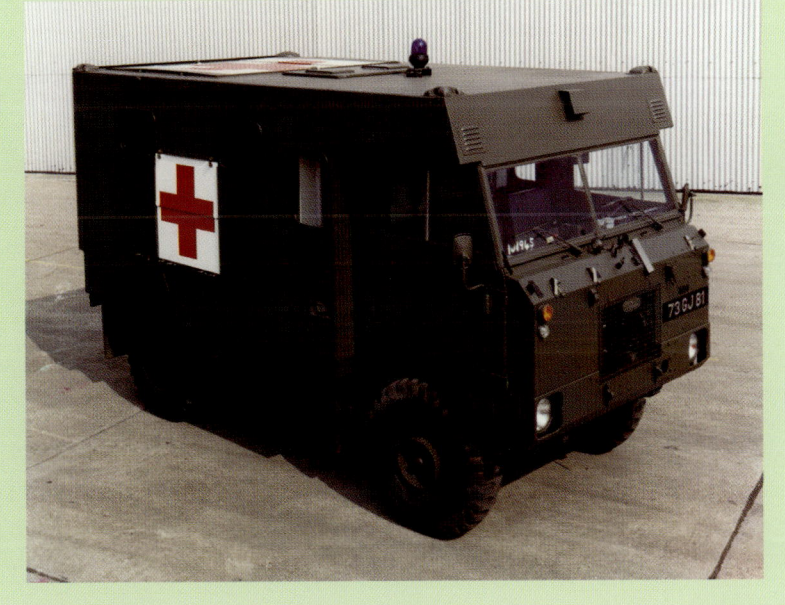

BRITISH 101 VARIANTS

ARMOURED 101

At about the time the first production 101s were entering service with the British armed forces, some work was done on creating an armoured variant. Sketches exist of a proposed armoured logistics vehicle, and the basic idea was then turned into a wooden mock-up. This design went no further, but in what may have been a related project, an early GS vehicle was armoured for evaluation at Chertsey.

The vehicle chosen was 956-00277A, which had been delivered to Hilton in July 1975 and had registration 62 FL 50. In November 1975, it went to 34 Central Workshops at Donnington, and the conversion was presumably done here. In January 1976 it went to Chertsey, and was photographed there in

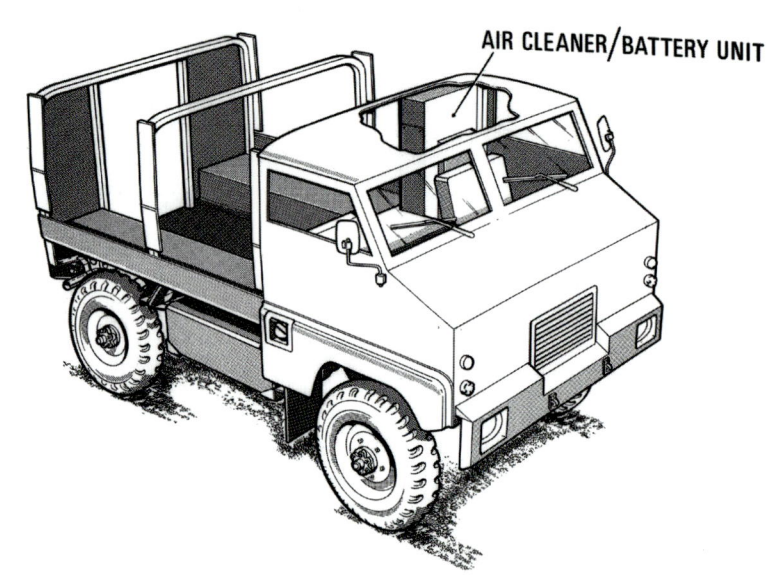

1 TONNE 4×4 ARMOURED LOGISTIC ROLE

FVRDE briefly pursued the idea of an armoured logistics vehicle on the 101 chassis, and their concept drawing was turned into a full-size wooden mock-up. However, nothing more appears to have come of the idea.
TANK MUSEUM

The idea of an armoured 101 did not go away, and in 1975, 62 FL 50 (956-00277A) was turned into an armoured vehicle by 34 Central Workshops. After evaluation at FVRDE, whose wing number is evident in this picture, it was returned to standard condition. Nothing more came of the idea. TANK MUSEUM

April 1976. The armour this time included covers over the lights and a hinged protective mesh for the windscreen. The vehicle's appearance suggests that it might have been evaluated for use in riot situations, such as the army was then encountering in Northern Ireland. However, no further examples followed. 62 FL 50 went back to 34 Central Workshops, was returned to standard and was issued to a unit as a GS vehicle.

Negatives in the Rover archives held at the Heritage Motor Centre show a 101 chassis as part of a set described as an 'armoured car' project. These date from February 1971, when the only chassis in existence would have been prototypes; the one pictured is likely to have been that for 101/FC/8C. No further evidence of such a project has emerged, and it may be that the negatives have been mis-filed.

CHALLENGER

The light weight and high power of the 101 made it well suited for work with Airborne Forces, and the REME workshop supporting the Airborne Brigade (10 Airborne Workshops) converted several for special REME roles.

Challenger – the code name was applied locally – was developed primarily to lift boxed stores and both vehicle and gun components. It may also have been used for tasks such as engine changes in the field. To create the Challenger, the 101's back body was stripped down and a small hydraulic crane with a one-tonne lifting capacity was fitted at the left-hand front of the load bed, where it was operated from the vehicle's centre power take-off. The maker of this crane is not clear, but it is thought to have been an Atlas type. Stabilizing legs were fitted underneath the body, and there was a small platform for the crane operator in the centre of the load bed.

As the tilt could not be fitted with this installation, REME manufactured a fully enclosed cab from metal and fibreglass, and added a mounting for the spare wheel on its roof. The front grilles had cutaway sections to accommodate an engineer's vice like that on the FRT vehicles (*see below*), but in practice no vice seems to have been fitted. It appears that the additional stresses placed on the chassis frame by the crane caused the frame to crack.

During Operation *Corporate* (the recovery of the Falkland Islands from Argentinean forces) in 1982, the main REME workshop that accompanied the British forces included one or more Challengers.

Research indicates that no more than three Challengers were built. The earliest was probably the one built on a pre-production 101; a photograph shows this with the name Challenger actually painted on the

BRITISH 101 VARIANTS

Challenger was a rare unit-converted version of the 101. The front view shows a vehicle after release into civilian life, and the two rear views show 69 FL 31, which is preserved in the REME Museum at Arborfield. AUTHOR

jib of the crane, but it is unfortunately not possible to identify which vehicle it was. The other two were constructed on 961-00104A (69 FL 11) and 961-00150A (69 FL 31), both 24V vehicles with right-hand drive; both survive. At the time of writing, 69 FL 31 was on display in the REME Museum at Arborfield.

DUAL-STEER 101

Prototype 101/FC/1 was mocked-up with steering wheels and instrument panels on both sides of the cab while at FVRDE in early 1970 (see Chapter 2), but the left-hand steering wheel appears not to have been connected. The exercise was probably intended only to identify potential problems with LHD vehicles, as no LHD prototype was available at the time.

Nevertheless, one production 101 was modified to have proper dual steering by an engineering company in Brighton that worked for the MoD. It was visually distinguished by an odd front end with an extension below the centre of the bumper, which gave protection to a transverse steering rod.

This vehicle was 68 FL 80 (961-00073A), which was attached to 4 RTR and was used for testing driver disorientation. The driver wore a helmet fitted with a screen and was able to see only what was projected onto the screen. The vehicle has now gone abroad, and is thought to be in a museum in Italy.

ELECTRONIC REPAIR WORKSHOP

REME units used a number of signals-bodied vehicles as electronic repair workshops, but the inventive minds of 10 Airborne Workshops also came up with a special variant of their own.

Just one vehicle was built, numbered 69 FL 21 (961-00140A). On top of the standard GS truck body, the workshops added a tall box-body with a Luton-type head extending over the cab and a large double door in the left-hand side. There was provision for an awning above this to give a covered working area outside the vehicle, and pairs of trailer legs were fitted front and rear to stabilize the vehicle when it was parked. The vehicle was probably used only for a short time before being replaced by a signals-bodied vehicle, 68 FL 53, that was supposedly taken from storage.

This unique 101 has sometimes been incorrectly identified as a power-generation vehicle.

69 FL 21 (961-00140A) was converted by a REME unit from a GS-bodied 24V truck into an electronic repair vehicle. UNKNOWN VIA DUNSFOLD DLR

BELOW: **The FACE fire control system is seen here in a 101 GS.** FREELANCE MILITARY WRITERS

BELOW: **The FRT recovery vehicle had a Tuckaway recovery crane mounted at the rear.** GEOFF FLETCHER

FACE VEHICLE

Several 101s were used to carry FACE – Field Artillery Computer Equipment. This artillery survey and fire control system was deployed alongside a gun battery. Typically, a gun battery that used 101 tractors would have its FACE in a 101 too; the vehicle appeared from the outside to be a standard GS type.

FACE was used in the British Army for a period of around 20 years from 1969. Its task was to compute target location and meteorological data to ensure accurate delivery of shells onto the target. Its main component was an Elliott Automation 920B computer that had to re-load its programme from a cassette of punched mylar plastic tape every time it was switched on. Other elements of the system were cabled to the computer inside the host vehicle.

FRT RECOVERY VEHICLE

Note that two different types of 101 were known as FRT (Forward Repair Team) vehicles. One was associated with the Rapier missile teams. The other was a special field recovery vehicle that was built by the REME unit of 10 Airborne Workshops.

It appears that four were built, beginning in the late 1980s; they were on 71 FL44 (959-00185A), 73 FL

00 (956-00142A), 73 FL 42 (956-00638A) and 55 KJ 70 (956-00269A). All were on winch-equipped chassis, and all retained their standard GS body configuration, the additional equipment being concealed under the standard canvas hood. They had a Dixon Bate Tuckaway vehicle recovery crane with a safe working load of 1.5 tons, which folded out over the rear cross member and could be swung back into the vehicle for stowage. In addition, an engineer's vice was mounted on the front bumper, set into a cutaway radiator grille. Although four were built, only two were in service at a time; the later two were replacements for the earlier pair, which cracked their chassis in the same way as the earlier Challenger vehicle developed by 10 Airborne Workshops.

GUN TRACTORS

Those 101s that were used in their original primary role as gun tractors were usually 956-series (RHD) or 959-series (LHD) 12V vehicles. Users included gun teams from the Royal Marines and the Parachute Regiment.

In theory, all gun tractors were equipped with a winch, and the accompanying limber tractor was a GS vehicle without a winch. In practice, however, vehicles without a winch were often used as gun tractors when the winch-equipped vehicle was unserviceable or no winch-equipped vehicle was available for issue from the RAOC Vehicle Depot.

These vehicles normally had inward-facing bench seats in the rear of the load bed, designed to accommodate two soldiers on each side. The front of the load bed would then be used for ammunition stowage and to carry each individual's personal kit.

MILAN CARRIER

MILAN was a portable, wire-guided medium-range anti-tank missile; the name is an acronym of its full French title: Missile d'Infanterie Léger ANtichar, which translates as light anti-tank infantry missile. In British Service, it was used by both army and Royal Marines units.

British MILAN teams regularly used Land Rovers of one type or another, and some 101s were equipped to meet their needs. These vehicles were 12V types, kitted out to carry six people, two firing posts and a rack of twelve re-load missiles. Access to the re-load rack was achieved by dropping the body sides and/or loosening the canvas hood.

MORTAR CARRIER

Some 12V 101s used by infantry battalions were specially adapted for the use of mortar platoons. On these vehicles, the forward half of the load area was divided into stowage compartments for ammunition and personal kit by vertical wire-mesh panels. Two of these stowage compartments were accessible from the outside by dropping the hinged body sides.

In the rear were three seats, bolted over the wheel wells. There was a two-person bench on the left-hand side and a single seat on the right, with a drop-down map table. The map table had rifle clips bolted to it for the rear occupants' weapons. There were then three containers fitted to the floor, which carried the mortar tube, base plate and tripod.

RAPIER TRACTORS

The 101 was designated as a tractor for the Rapier missile system very early on in its military career. The first Rapier tractors were delivered to the British Army in January 1975, and deliveries continued at a fairly constant rate until mid-1977. The vehicles were based on standard production 101s and in the beginning were converted to their designated roles by the British Aircraft Corporation (BAC) at Stevenage before onward delivery to their end-users. Some later vehicles – notably those used by RAF 6 Wing – were converted by 18 Command Workshops at Bovingdon.

The Rapier was (and remains) a highly mobile surface-to-air missile system that was split between two trailers. One trailer was the missile launcher, and the other carried the associated optical tracking radar, which from 1979 was supplemented by a Blindfire radar system. The tracking radar used an IFF system, which swiftly distinguished potential targets from friendly aircraft. The Rapier operator could choose between manual or fully automatic firing; in the latter case, the system launched a missile as soon as it detected an enemy aircraft.

LEFT: **Reconstructed for the War and Peace Show at Beltring in 2000, this tableau shows a Rapier system ready for use. Behind the 101 is a trailer containing additional missiles. The 101 itself, 00 AM 95, has chassis number 956-1062A and is one of the 12V vehicles converted to 24V specification for 6 Wing RAF. It originally belonged to 19 Squadron, based at RAF Brize Norton.** AUTHOR

As they really were: this LHD 101 was pictured with its Rapier trailer while serving with the Royal Artillery in BAOR. BAOR PUBLIC RELATIONS

Developed by BAC, the Rapier system began to enter service with British forces in 1975 (although official propaganda has suggested it was in service as early as 1971). It had replaced most of the earlier systems by 1977, and at the time of writing was expected to remain in service until at least 2020.

The main function of a Rapier detachment was to defend military units against low-altitude air strikes, and the system's ready mobility made it ideal for the defence of temporary airfields and static units operating in the field, as well as more permanent installations. Typical deployment time from arrival on site to operational readiness was around 30 minutes.

There were two types of 101 Rapier tractor. The vehicle designated to tow the missile trailer was known as the FUT (firing unit tractor), and was based on a 24V 101. It carried a three-person crew, the third person being a radio operator whose equipment

These two pictures originate with the Australian military, although UK Rapier vehicles were the same. They show the layout of the launch vehicle tractor, which the Australians called the Firing Unit Tractor or FUT. REMLR

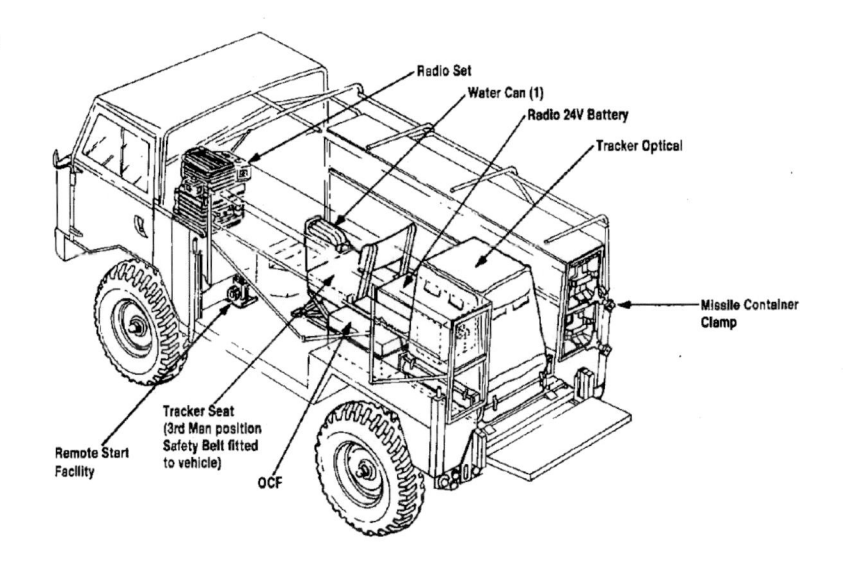

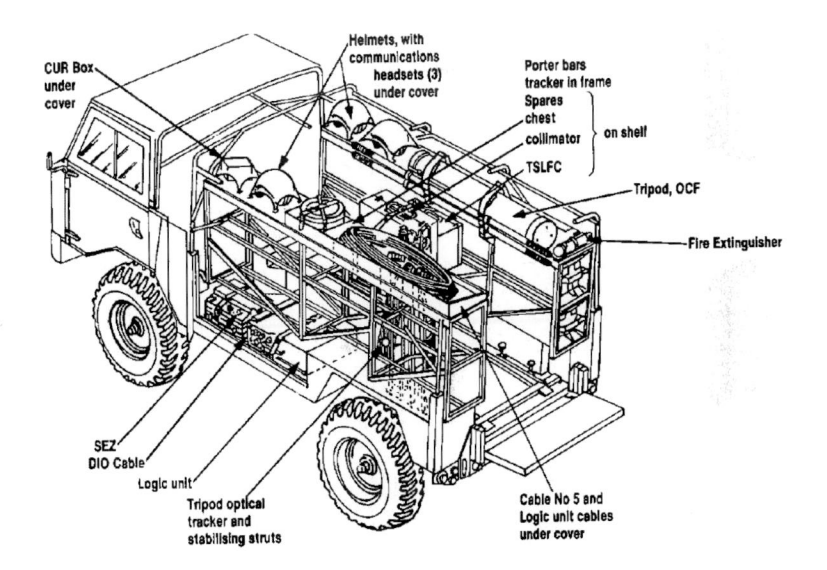

demanded the 24V chassis specification. The one designated to tow the tracking radar system was known as the TRT (tracking radar tractor), was based on a 12V vehicle and carried a two-person crew.

Both types of vehicle were fitted with special racks to carry four missiles. These racks were secured to a false floor in the load-bed, and other equipment needed by the Rapier team was distributed between the two vehicles in a detachment. From the outside, the vehicles looked like ordinary GS trucks because the missile racks and other equipment were completely concealed under the standard 101 canvas hood. Rapier tractors always had a winch, and vehicles selected for later conversion to the Rapier tractor role were fitted retrospectively with winches if they did not already have them.

In British service, a Rapier detachment generally consisted of three Land Rovers. The third vehicle (initially a Series III 109 Land Rover, latterly a One-Ten) was the stores vehicle, which carried the crew's

106 ■ BRITISH 101 VARIANTS

Again Australian in origin, these pictures show the layout of the Rapier TRT (Tracking Radar Tractor). The Australians actually called this an RTV (Radar Tracker Vehicle). REMLR

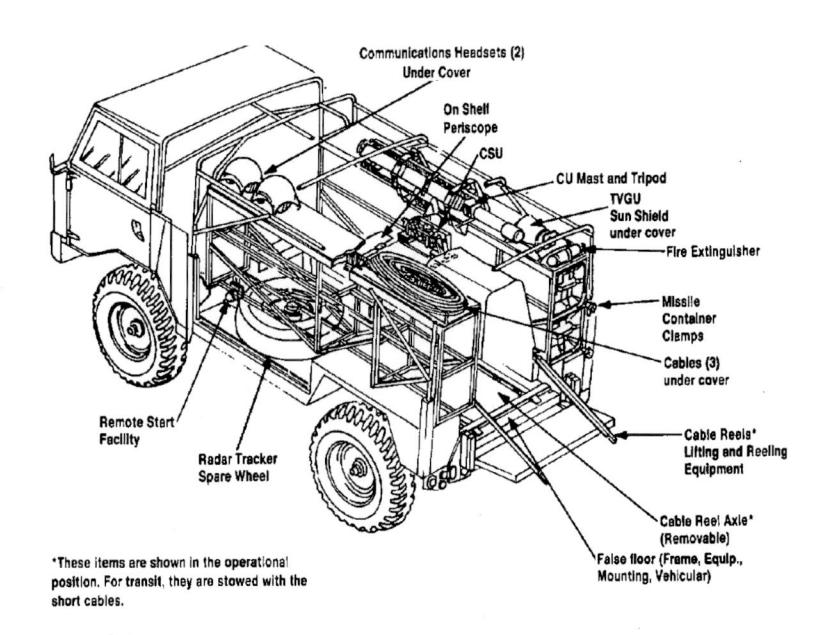

kit. This third vehicle was intended to tow an FV2411 re-supply trailer with a further nine re-load missiles.

In the broadest sense, there were two end-users of the Rapier tractors. These were the Royal Artillery and the RAF Regiment. The Royal Artillery most commonly used their Rapiers to defend static military units in the field. The RAF Regiment used theirs to provide cover for temporary airfields (such as those used by the RAF's Harrier jets), may also have used them for a while to defend the permanent air bases at Brüggen, Laarbruch and Wildenrath in Germany, and also for the defence of permanent USAF air bases in the UK.

The unit charged with this latter responsibility was 6 Wing, which was established on 1 July 1983 specifically for the purpose. Its 101s carried RAF

BRITISH 101 VARIANTS ■ 107

RIGHT: **This is the frame on which the components of the Rapier system were mounted in the FUT vehicle. The extension furthest from the camera carried the seat for the third crew member, the radio operator.** 101FCC&R

BELOW: **For comparison, this is the frame from the TRT vehicle.** 101FCC&R

ABOVE: **Both these pictures are also from Australia, and they show the Rapier trailer as it was when towed, and as it was when set up ready for operational use.** REMLR

RIGHT: **The generator used with the radar tracking system was a free-standing unit, but was transported on the radar trailer.** 101FCC&R

108 ■ BRITISH 101 VARIANTS

The TRT, in this case an RAF example, is seen here loaded with missiles in its racks. 101FCC&R

The RAF 101 on the left is a TRT with its radar trailer, while that on the right is an FUT with its launcher trailer. 101FCC&R

BRITISH 101 VARIANTS 109

The tall tilt identifies 73 FL 77 (956-00749A) as a Rapier test vehicle. 101FCC&R

formed in mid-1984, and covered the air bases at Fairford and Upper Heyford; 20 Sqn was based at Honington, was formed in mid-1985, and looked after Alconbury and Mildenhall; the third squadron was 66 Sqn, based at West Raynham and was actually the earliest to be formed, in November 1983 – this had responsibility for Bentwaters and Woodbridge air bases. In addition, the USAF funded a Rapier Training Unit Flight.

The 6 Wing Rapier tractors fell into two groups of thirty-five vehicles each, which were:

00 AM 72 to 01 AM 06   24V FUT vehicles
01 AM 07 to 01 AM 41   12V TRT vehicles

All seventy vehicles were on 956-series chassis, which had originally been built with 12V electrical systems and were taken out of storage. Of these, it appears that sixty-nine were 'left-overs' from the ambulance contract that were fitted with 'pattern' GS bodies by Marshall's. The thirty-five FUT vehicles were converted to 24V in 1983–84 before entering service, and the work was almost certainly done on a special assembly line at the Land Rover factory in Solihull. This was set up in a building next to the old service department that was known as the IFV

registration numbers and were crewed by RAF personnel, but its three squadrons were actually funded by the USAF: 19 Sqn was based at Brize Norton, was

building. Those letters probably stood for 'interference, fighting vehicles', because the building was the place where the production department had earlier carried out end-of-line screening tests on FFR vehicles. It was a narrow building with an inspection pit that ran down its full length and was invaluable when work had to be done underneath the vehicles.

Rapier missiles were first used in anger during the Falklands campaign in 1982, with mixed success. Rapier 101s were also used during the first Gulf conflict in 1990–91, even though the Land Rover 127s, which were scheduled to replace them, had already been delivered and were theoretically available for service. In most cases, these deployments required defence of a temporary airfield, and in these cases a Rapier detachment would drive to its launcher location, set up the launcher and tracker trailers (often with the aid of the 101's Nokken winch), and then remove the vehicles from the scene to reduce the profile of the missile battery.

The Rapier missile tractors were supported by a number of Rapier test vehicles, all of them based on LHD 12V chassis. These were used by the Forward Repair Teams and were sometimes known by the same FRT acronym as the special recovery vehicles built by 10 Airborne Workshops. The army had twenty-six of them, all converted from 24V GS vehicles.

The Rapier test vehicles were recognizable by a stepped canvas hood, where the section over the load area was higher than that over the cab. Under that stepped hood was the automatic test equipment (ATE) for the Rapier system. Each of the army's three Rapier Regiment REME Light Aid Detachments (LADs) had six vehicles, two for each Forward Repair Team (FRT) attached to each battery. Seven more vehicles were issued to the engineering flights of RAF Rapier Squadrons. These figures suggest that just one vehicle was kept in reserve.

SIGNALS BODY

A box-body version of the 101 had been in the plan since very early in the 101 project, and the FVRDE internal brochure that was probably produced in late 1968 shows that it was expected to be suitable for workshop, command post, and signals uses. The first box-body built was probably that constructed by

THE 101 RAPIER TRACTORS

It is not certain how many 101s were used as Rapier tractors, but a careful trawl through records gives a provisional figure of 272, made up as follows:

Royal Artillery (provisional total 122, FL-series registrations)
7	956-series RHD 12V radar trailer tractors
4	959-series LHD 12V radar trailer tractors
38	961-series RHD 24V launch trailer tractors
73	964-series LHD 24V launch trailer tractors

RAF Regiment (provisional total 80, AM-series registrations)
8	956-series RHD 12V radar trailer tractors
29	959-series LHD 12V radar trailer tractors
9	961-series RHD 24V launch trailer tractors
34	964-series LHD 24V launch trailer tractors

RAF 6 Wing (total 70)
35	956-series RHD 12V radar trailer tractors
35	956-series RHD 12V launch trailer tractors (all converted to 24V)

Marshall's on prototype 101/FC/6 (see Chapter 2), but a further prototype was then built in mid-1974 by Cammell Laird. This was on one of the pre-production vehicles, 961-00010A (68 FL 44), and at Land Rover it was known as the Rebev vehicle. Exactly what that meant is unclear; perhaps it was simply a misunderstanding of the military term Rebro, which meant a rebroadcast vehicle.

Nevertheless, the Cammell Laird box-body design was not adopted for production; instead, the MoD turned to Marshall's of Cambridge for the volume contract. That volume was not large. A best guess is that there were around 100 signals-bodied 101s. A total of 99 can be reached by adding these figures:

26 for the Army with LHD
41 for the Army with RHD
7 for the RAF (all LHD)
25 Intercept Complex vehicles (all LHD)

The Cammell Laird prototype was modified to be more similar to the production vehicles, and adding this to the total above would make exactly 100 examples.

Though generally known as the signals body or radio body, the Marshall's box-body was not intended exclusively for signals use. In fact at least six of the army's vehicles and all seven of those for the RAF were on 12V chassis. Although some of the 24V vehicles certainly did become communications vehicles, many appear to have been used by REME units for radio and electronics repair duties. Others were used as Intercept Complex vehicles with the Vampire electronic warfare units. A few saw service as command posts.

During the First Gulf War (Operation GRANBY), some carried air-conditioning packs in the roof box. Nine of the vehicles deployed on that operation, including 71 FL 71 (964-00012A) and 72 FL 56

A first prototype of the signals body was constructed by Cammell Laird on a pre-production chassis, 964-00006A (68 FL 44), but this design was not pursued for the production batch. TANK MUSEUM

112 ■ BRITISH 101 VARIANTS

This was the production design of signals body, built by Marshall's of Cambridge. The vehicle is 72 FL 06 (964-00279A), originally delivered from Rover in 1976 and now in preservation.
AUTHOR

BRITISH 101 VARIANTS ■ 113

(964-00134A), became part of 1 Field Laboratory Unit (1FLU), which used an experimental biological detection system (BDS) that monitored for the presence of biological warfare agents. The vehicles were converted by the CBD Porton Down Engineering Section between 10 December 1990 and 6 January 1991, when they received detection sub-systems, air-conditioning, refrigeration and radios. While on Operation GRANBY, they were on charge to the Chemical Defence Establishment at Porton Down.

The signals' body was characterized by a low, flat roof, which normally carried a box-like structure with two semi-circular channels that could be used like a roof rack to carry such items as aerial masts. There was a fixed ladder on the right-hand side at the rear to give access to the roof stowage, and a large tail door with a sliding window. The right side of the body was windowless, but the left side had double doors towards the front, with a fixed window in the rearward one. One more window, this time a sliding one, was positioned towards the rear of the left-hand body side. The body plate applied by Marshall's identifies the maker as MCE (Marshall Cambridge Engineering), and carries the description of 'body utility' and the military code FV773705.

Some vehicles had ports and sockets in the body sides for specialist equipment, and some had provision for a penthouse to be fitted to the side to extend the working space. All probably had the triple-jerrycan rack on the left of the chassis, and some may

ABOVE: **Reconstructed as faithfully as possible, this is the interior of 72 FL 06 in preservation.** AUTHOR

RIGHT: **The Marshall's body plate was on the outside of the signals body.** DUNSFOLD DLR

ABOVE: **Vehicles were sometimes adapted in the field for specific roles. This signals-bodied vehicle was fitted with large roof panniers by the REME unit that used it. Seen here after sale into the civilian market, it was originally 69 FL 23 (961-00142A).** DUNSFOLD DLR

OPPOSITE PAGE: **Some of the signals-bodied 101s were used as elements in the Intercept Complex, which was intended to intercept enemy battlefield communications. This is 72 FL 70 (964-00148A), with its specially-adapted Sankey trailer that carried a Clark extending mast.** TANK MUSEUM

have carried a spare wheel on the roof above the cab in service, although this was not a standard arrangement. At least one vehicle (69 FL 23, 961-00142A) was modified in service by REME to carry two large boxes on the roof.

The Intercept Complex vehicle was probably first tested in 1982. It was visually indistinguishable from other signals-bodied 101s, although most carried the spare wheel mounted on the front panel to make more room for equipment in the body. A one-ton Sankey trailer was towed behind, modified by Clark Masts to carry a large mast aerial and an aluminium alloy ladder that was needed when erecting or dismantling the aerial. These vehicles replaced the Series III 109s that had been used for the Intercept Complex until the early 1980s. The trailers were probably the ones that had formerly been used with the 109s; the one trialled with 72 FL 70, for example, had registration number 16 EN 79, which dates it to a 1964–65 order.

There appear to have been twenty-five Intercept Complex vehicles, all on LHD 964-series 24V chassis. These were:

70 FL 73 to 70 FL 74
70 FL 95
70 FL 99
71 FL 01

BRITISH 101 VARIANTS 115

BRITISH 101 VARIANTS

71 FL 03
71 FL 09
71 FL 22
71 FL 24
71 FL 30
71 FL 34
71 FL 72
71 FL 81
71 FL 98 to 72 FL 01
72 FL 19
72 FL 29
72 FL 32
72 FL 39
72 FL 63
72 FL 71 to 72 FL 72
72 FL 77

Note that the box-body used for the Vampire 101s had a completely different design.

VAMPIRE

The Vampire was a special hard-body version of the 101 that was used for electronic warfare purposes. The name is actually an acronym, which decodes as follows:

V = vehicle
A = army
M = mobile
P = position
I = interferometry
R = radio
E = equipment

Briefly, the Vampire was designed to operate close to the FEBA (forward edge, battle area) where its task was to intercept enemy battlefield communications. It then encrypted those communications and relayed them to an Intercept Complex that had been

The Vampire was a top secret signals intercept vehicle. 75 FL 98 (964-00100A) may have been an early prototype. Note the roof-mounted mast and the tall and rather unstable-looking trailer, which appears not to have been standard issue for these vehicles.
PLESSEY VIA DUNSFOLD DLR

established further back, typically 5km behind the FEBA. The Complex then analysed the communications and sent appropriate information to headquarters. The Intercept Complex itself consisted of five or six vehicles and was originally based on a Series III 109in Land Rover, but in the early 1980s was transferred to a rare variant of the 101 signals body.

The Vampire system was developed in the late 1970s as a replacement for the German-made Telegon 4 system and was late entering service. The first units reached 14 Signal Regiment, the electronic warfare regiment attached to the 1st British Corps of the BAOR, in November 1983. The Regiment was then garrisoned at Celle in West Germany.

The prototype was built on 964-0004A, which carried registration number 54 BT 07. The basic box-bodywork on this and on the production batch was constructed by Marshall's of Cambridge. It had a lower roofline than that company's contemporary ambulance bodies for the 101, with a single access door (including a window) in the right-hand side and a ladder at the rear that gave access to the roof. Some Vampires, but probably not all, were fitted with a Sperry Navigation System, characterized by a long arm that projected over the front of the vehicle above the cab. An early version of this had been tried out on prototype 101/FC/6 (see Chapter 2).

A best guess is that eighteen Vampires (plus one prototype) were built. The first one was delivered in September 1983, and research suggests that each vehicle cost over £440,000 when new. As each vehicle also towed a specially equipped trailer, the total cost per vehicle must have been considerably higher. All the vehicles appear to have been built on left-hand drive 24V chassis, probably taken from the store of unissued vehicles at Ashchurch.

After bodying at Marshall's, the vehicles were then delivered to Plessey Avionics, who fitted them with direction finding (DF) receivers of their own manufacture. The aerials for the DF receivers were attached to a pneumatically-operated Clark elevating SCAM mast, located in a hinged mounting at the rear of the roof. When travelling, the mast was carried diagonally across the roof, but when the vehicle was operational, it was raised into a vertical position and its lower section was secured to the vertical rear panel. When fully extended, the mast reached a height of 21m (69ft) from the ground, and carried a Narrow Aperture Adcocks array. This was a cross-piece receiver aerial with a total of eight dipoles, two on the end of each 'arm' of the aerial. The mast was normally secured by cables, but would often have additional protection from wind if the Vampire unit was deployed in woodland. Further stability came from levelling jacks on the rear corners of the vehicle.

All this produced an extremely heavy vehicle, with a total weight of around 3.5 tonnes. The Vampires also had a high centre of gravity – located just below

This interior photo is of 75 FL 93 (964-00358A), probably the most complete example of a Vampire in preservation. Carefully restored, it gives a good impression of the working conditions for the operators.
GEOFF FLETCHER

the driver's shoulder level – and were known to keel over sideways on occasion. The Nokken winch that was standard equipment must have come in handy if pressed. Other accidents were not unknown; the vehicles would usually be deployed away from roads and in the dark, and one story tells of a Vampire that ripped off both its axles when it hit a pile of logs in the dark.

Each vehicle towed a specially equipped Sankey trailer. One early Vampire (75 FL 98, 964-00100A) used a standard one-ton trailer (19 GJ 43) that had been equipped with a tall and heavy looking box-body. However, this may have been unique; subsequent vehicles seem to have used more stable-looking standard open trailers fitted with a number of tailor-made boxes.

The trailer carried a wide aperture ground array aerial system and when loaded it added another 1.1 tonnes or more to the gross train weight of the Vampire system. The ground array consisted of three free-standing monopole aerials that would be deployed in the form of an Isosceles triangle, together with a test antenna that would be located in the centre of the triangle. Its functions were to increase the accuracy of the Vampire system and to enhance the frequency range that was being searched. However, it was not always possible to use the ground array operationally, as it required an area about one-third the size of a football pitch for full deployment. When used, it was set up between 100 and 150m (328–492ft) away from the Vampire vehicle and was linked to it by 32-core cables that were normally carried in the trailer.

The interior of the Vampire body was specially prepared for its task, with fully insulated walls lined with aluminium and painted light green. The body was arranged to be light tight, and the small window in the side door had its own sliding cover. To provide heating for the body, there was a small petrol-powered heater mounted just behind the driver's seat, and there was also an air-conditioning unit in a grey-painted box.

In 14 Signal Regiment, there were four Vampire Troops, with four Vampire vehicles in each. The regiment probably kept two more vehicles in reserve. Each Vampire was supported in the field by a Series III 109in Land Rover that towed a standard Sankey trailer. This vehicle and trailer carried all the kit belonging to the Vampire crew, including their 9×9 tent. The four Vampire units would be deployed in a line (usually known as the 'baseline'), around 1 to 2km (½–1¼ miles) from the FEBA (forward edge battle area) and with around 3 to 5km (1¾–3 miles) between each unit. So deployed, they had a range of around 50km (31 miles), line of sight dependent.

During the Cold War, the Vampires seem to have spent most of their operational life with 14 Signals Regiment in Germany. They would have been regularly deployed on exercises to train for their wartime

Padlocks were used on the doors of Vampires to improve operational security. This one was on the first prototype, 964-00004A. AUTHOR

75 FL 93 is seen here with the vehicle-mounted mast partially deployed (it could be raised to a much greater height) and with a trailer that was probably typical of those used operationally. AUTHOR

role. However, once the Berlin Wall came down and there was a thaw in relations between East and West, their usefulness was reduced. Some Vampires lost their electronic warfare equipment at the start of the First Gulf War, when it was transferred to FV432 Armoured Personnel Carriers for use in the desert. Supposedly, four or six FV432s retained this equipment until the late 1990s.

At least four and possibly five or more Vampires nevertheless remained in service for several more years; among them was 75 FL 93 (now the most complete surviving example), which served in its original role with 237 Signal Squadron after the squadron relocated from Germany to the UK in April 1993. The final example is thought to have been 'cast' on 29 February 2000.

WINTERIZED AND WATERPROOFED

A winterization kit, also known as a cold weather kit, was developed early on and was fitted to those 101s used in cold climates. It was developed on one of the pre-production vehicles (956-00002A), and a feature of this early system was a small metal radiator header tank mounted externally ahead of the windscreen.

The cold weather kit was developed to meet a specification laid down in SR 01-00/09 section 3, paragraph 5(a). It used a petrol-fired boiler made by Willco Engineering of Aldershot and known to them as a thermocyclic heater. The same boiler was used in cold weather kits for the 88in and 109in Land Rovers, and also the Bedford MK four-ton truck. Early prototypes relied on the thermo-cycle principle for water circulation, and the first 101 installation used a MkI boiler. A switch was soon made to the MkIV type, and later still the carburettor was changed from an Amal type to one made by Willco themselves.

The first installation was done by Rover engineers on prototype 101/FC/3 in April 1970, and the second in January 1972 on 101/FC/8C. Tests were done in the cold room at Smith's in Witney, and as a result an improved version was drawn up. This included two eyeball vents to improve heat distribution, and these were fitted to vehicle P5 (964-00005A), which was one of the twenty-five pre-production vehicles for the MoD, who subsequently tested it at Chertsey. In

> ### THE VAMPIRES
>
> It is difficult to identify the Vampires in military records, not least because some of the vehicle cards were withheld for many years. Even at the time of writing, not all were held at Deepcut RLC Museum with the other army vehicle records.
>
> The probable list of Vampires is as follows:
>
> | | |
> |---|---|
> | 54 BT 07 | (total 1) – prototype, possibly not operational |
> | 75 FL 41 to 75 FL 42 | (total 2) |
> | 75 FL 44 | (total 1) |
> | 75 FL 60 | (total 1) |
> | 75 FL 76 to 75 FL 77 | (total 2) |
> | 75 FL 90 to 75 FL 93 | (total 4) |
> | 75 FL 95 | (total 1) |
> | 75 FL 97 to 75 FL 98 | (total 2) |
> | 76 FL 21 | (total 1) |
> | 76 FL 29 to 76 FL 30 | (total 2) |
> | 76 FL 33 | (total 1) |
> | 76 FL 49 | (total 1) |
> | GRAND TOTAL | 19 |

addition, a version of the winterization kit was tried out on 956-00002A (CXC 347K), and both this and 03 SP 72 (959-00002A) had the feature of a small metal radiator header tank mounted externally on the front panel ahead of the windscreen. A winterization kit also appears to have been fitted to BXC 676K (964-00001A) in August or September 1972.

A semi-permanent Waterproofing installation was also developed, and this became available for issue to units in May 1980. Several 101s were allocated to ATTURM (Amphibious Trials and Training Unit, Royal Marines) at Instow in north Devon, and were used for development and trials of this installation. The waterproofing kit was regularly found on vehicles allocated to Royal Marines' units, and 29 Commando Regiment RA was one unit that employed waterproofed 101s as gun tractors. The vehicles were used for deep-water wading during amphibious landings, and were also carried under Sea King helicopters for air-mobile operations.

CHAPTER SEVEN

THE 101 IN OVERSEAS SERVICE

As Chapter 4 explains, not every 101 was delivered to the British armed forces. Of the 2,666 built, around 270 and possibly more than 300 were sold overseas. The exact figure is in dispute because records are either vague or non-existent. Nevertheless, as a working estimate, around 10 per cent of production went to the armed forces of foreign nations.

This chapter aims to give an idea of what happened to the 101s that were sold outside the UK, and to record the identities of those vehicles.

AUSTRALIA

During 1968, the Australian Army called for manufacturers to submit proposals to meet what it called DDX 188, a draft development specification for a one-ton truck. The initial Australian Army mock-up vehicle, called XF4, had in fact used a number of Land Rover parts, and of course the Australian Army had used Land Rovers in large quantities since the 1950s. Most in service at the time had actually been built in Australia from CKD kits shipped out from Solihull, and they incorporated a substantial local content.

The design concept was for a forward-control 4×4 with a body that could be converted from tray to flat bed. The vehicle was to have differential locks on both axles and power take-off (PTO) arrangements for both ancillary equipment and a powered trailer. It was to use commercially available components wherever possible. Chrysler, Ford, GM-Holden, International Harvester, Rover and Volvo all responded with tenders.

Rover's tender for the contract seems to have embraced four different options. Tender A was a 109 one-ton production model; Tender B was a standard (as envisaged at that stage) 101 Forward Control; Tender C was a widened 101 Forward Control; and Tender D was an outline concept that looked very much like the Australians' own XF4. Both Tender C and Tender D were described as meeting DDX 188 requirements, and Tender C would have seen the 101 widened from 72in to 86in (183cm to 218cm), with a corresponding increase in track from the existing 60in to 70in (152cm to 178cm).

No vehicle was ever built to the Tender C or Tender D specifications, but Rover did persuade the Australians to look at two vehicles. One was a standard 101 prototype with the 3-litre 6-cylinder engine then in use, and the other was a specially built prototype with the Australian-built 3.6-litre Ford Falcon 6-cylinder engine. (There are further details in Chapter 2.) Some trials were carried out over the spring of 1969 with these, of which at least one was known to the Australians as a 'prototype DDR narrow version'.

However, the Australians were not impressed. Only Ford and International Harvester were invited to supply trials vehicles, and in 1971–72 these were examined alongside the domestic XF4 vehicles. Eventually, the project was scrapped altogether.

Despite this early rejection of the 101, the Australians did eventually end up with a sizeable fleet of them. Their requirement was not for the 101s as such, but for the Rapier missile system that was manufactured by the British Aircraft Corporation. BAC were presumably asked for recommendations about towing vehicles and, as the company was already manufacturing a 'fit' for the 101, must have

THE 101 IN OVERSEAS SERVICE

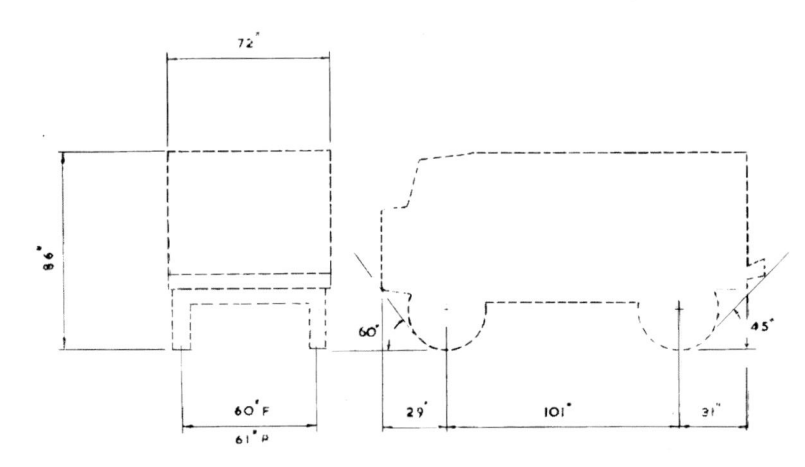

TENDER 'B'   LANDROVER 101˝WB FORWARD CONTROL

These sketches come from the unsuccessful attempt to interest the Australian military in the 101. Tender B is the standard vehicle, and Tender C illustrates a widened version (which was never actually built). The profile makes clear that the drawings date from the time of the 'snub-nosed' 6-cylinder prototypes in the late 1960s. LAND ROVER AUSTRALIA

TENDER 'C'   LANDROVER 101 WB FORWARD CONTROL   (WIDENED TO DDX 188 REQUIREMENTS)

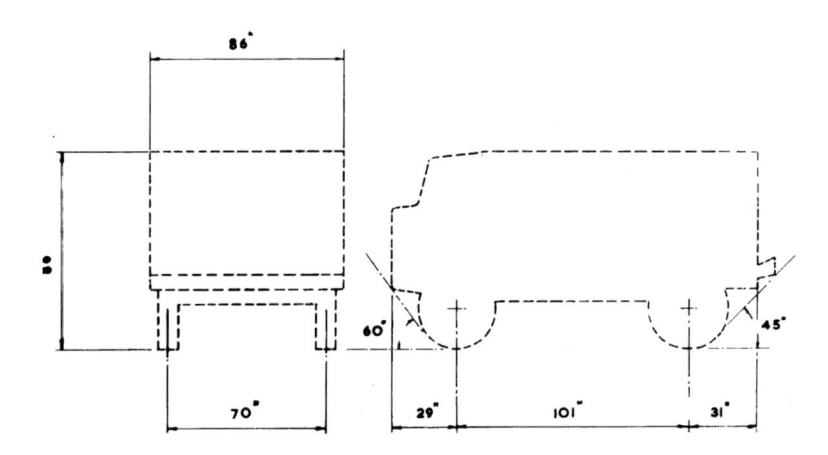

recommended the vehicle to the Australians. So the 101s came as part of the Rapier missile package, and were supplied not by Land Rover directly, but by BAC.

This was an unusual way of doing things. As BAC supplied the vehicles direct to the Australian Defence Force, the Land Rover national sales company in Australia was not involved. It was therefore somewhat surprised to find that the Australian Army expected the same level of logistical and maintenance support for these vehicles as was already being supplied for the normal-control Land Rovers it had in service.

With very little information available, the Land Rover team at Leyland Australia (as it was then known) scrambled to help, and quickly got up to speed to provide the support that was expected.

The delivery of fifty RHD 24V 101s as Rapier missile tractors was preceded by the delivery of two 961-series chassis to UK RHD specification for acceptance trials in 1977. These were drawn from UK MoD stock on 14 April 1977 and were sent to BAC at Stevenage; their original numbers were 68 FL 55 (961-00048A) and 68 FL 57 (961-00050A). For the trials in Australia, they carried military registrations 29-170 and 29-171. After the trials, both were returned to the UK and entered UK military service with their original registration numbers. The main delivery of fifty vehicles arrived during 1978, in two batches, and all were numbered in the 962-series reserved for 24V RHD export vehicles.

The first batch of thirty went to British Aerospace at Stevenage between 14 September and 16 November 1977, and the second batch of twenty between 1 June and 9 August 1978. These vehicles all had Nokken winches, and all were painted in Olive Drab semi-gloss paint (which has Land Rover code number LRC 286). They were on chassis numbers 962-00035A to 962-00064A and 962-00073A to 962-00092A.

These vehicles received Australian military registrations 29-405 to 29-426, and 30-238 to 30-263. They were not registered in chassis number order. Records at the UK end make quite clear that fifty vehicles went to Australia, but it has so far proved impossible to find more than forty-eight of them in Australian military records. The 'missing' vehicles are 962-00078A and 962-00090A, both from the second batch.

The Australian Rapier tractors were initially delivered to 41 Support Battalion, which is the ordinance and supply unit for South Australia. This was in 1978.

One of the two Australian trials' vehicles, wearing registration plate 29-170, is put through its paces at the Monegeeta proving ground. Braking was obviously impressive, if a little nerve-wracking for the driver! ADF VIA REMLR

124 ■ THE 101 IN OVERSEAS SERVICE

These official photographs of a Rapier tractor and trailer when new show how the re-load missiles were stored in the body of the vehicle. They would normally have been concealed by the canvas hood. ADF VIA REMLR

After allocation of army registration numbers, they were then shipped to user units.

The primary user was 16 Air Defence Regiment, based at Woodside Barracks in South Australia and part of the Royal Regiment of Australian Artillery. However, the point of the Rapiers was to create a highly mobile air-defence system, and 16 ADR's responsibilities in wartime included the protection of army units in the field, defence of the Royal Australian Navy's support ships and defence of Royal Australian Air Force airfields. So the 101s went wherever the Rapiers were needed – which, over the next dozen years or so, meant on exercises all over Australia.

The full breakdown of the 101 allocation has not yet been discovered, although the Register of Ex-Military Land Rovers in Australia has done some sterling research in this area. So far, it looks as if twenty-eight vehicles were allocated to 16 ADR, while a handful went to training centres. There are thought to have been as many as four at the RAEME mechanics training centre, and maybe two at the RACT driver training school. Both of these were in Bandiana, Victoria. That allocation would have left a dozen or more in reserve, which seems quite a high number but is not unreasonable.

Much as in the British Army, the Rapier sections each operated with three vehicles, all of which seem to have been 101s. The first 101 towed the Rapier launcher, the second towed the tracking radar unit and the third had the resupply role and carried spare missiles; British practice was to allocate this task to a normal-control Series III 109 Land Rover.

The 101s remained in service until the start of the 1990s, when they were replaced as Rapier tractors by air-defence derivatives of the Perentie 6×6 Land Rovers. Stripped of their Rapier equipment, they were then sold off to the civilian market. Most went through a pair of auctions in Adelaide, although others were auctioned direct from army bases. A number ended up in enthusiast hands.

Not all of the Australian 101s made it into civilian life. At least three were written off while in service: one hit a bus head-on; one was burned out when somebody tossed a lighted cigarette into it, only to discover that it had a fuel leak; and a third disappeared into a sinkhole on the Nullarbor Plain, from where it was not (and presumably could not be) recovered.

As for the Rapier missiles, they remained in Australian military service for a few more years, until the air defence 6×6s were converted to carry the later RBS 70 missiles.

Australian Rapier tractors and trailers in transit by rail.
NEIL DAILEY/REMLR

126 ■ THE 101 IN OVERSEAS SERVICE

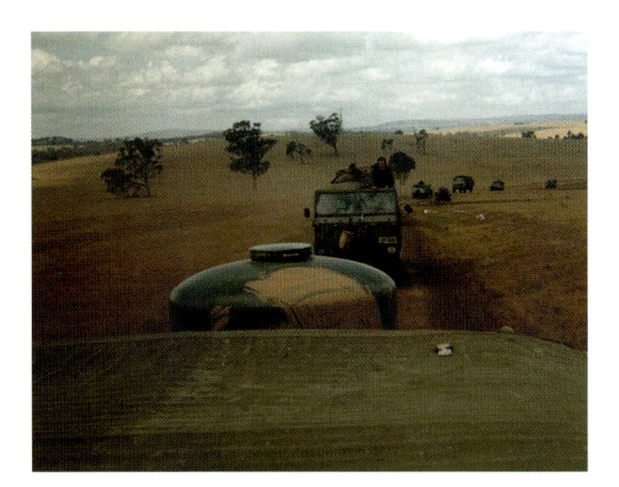

Australian Rapier tractors in convoy on a deployment; in this case, not all of them seem to have been towing their Rapier trailers. NEIL DAILEY/REMLR

The Australian 101s were sold off at auction; this line-up was pictured at a disposal yard in Adelaide in 1991. ROD GENN/REMLR

BRUNEI

A total of forty RHD 101s was delivered to Brunei during 1977. Six of these were 12V types (957-00007A to 957-00012A), which went to the Crown Agents in Brunei between August and October that year. These were followed by thirty-four vehicles, which had 24V electrical systems (962-00001A to 962-00034A). No further details of these vehicles are currently available.

DUBAI

Dubai took thirteen LHD 12V 101s during 1975–76 on the recommendation of Robin Wimbush, who was then the Land Rover representative in the Middle East. These were on chassis numbers:

959-00027A
959-00031A
959-00037A
959-00046A
959-00060A to 959-00067A
959-00281A

The first four were delivered in December 1975 and the next eight in February and March 1976. The final example, 959-00281A, was despatched through the Crown Agents in Dubai in November 1976.

Dubai had originally ordered at least three more 101s. Three vehicles built for Dubai, which became cancelled orders, were sold in approximately August 1977 to Airways Garage at Hayes in Middlesex (which was then owned by well-known Land Rover enthusiast Richard Beddall). One of them, 959-00282A, which follows on from the last of the vehicles actually delivered, was sold to the Venturers Search and Rescue organization in Lyndhurst. It had been re-worked before sale by Land Rover. The identities of the other two vehicles, which were probably sold on to export dealers, have not yet been established.

One of the Dubai vehicles (959-00046A) was brought back to the UK in the later 1980s by a civilian owner. The back body had been fitted with an expanded metal cage, which had an entry gate at the rear, and the whole cage was completely concealed under the canvas tilt. It may have been used for transporting prisoners during its military career, but there has been no confirmation of this.

EGYPT

The Egyptians appear to have bought three 101s in 1976–77 as Swingfire (Beeswing) missile launchers. They were supplied through BAC, makers of the Swingfire system. The vehicles remain unidentified, but it is likely that they were drawn from the batch of six 959-series 101s (959-000135A to 139A) whose end-users remain unclear. Alternatively, they may have been 964-series 24V vehicles. If the Egyptian vehicles did come from the 959-series batch, the possibility remains that there were six Egyptian 101s and not just three.

The Egyptians took 101s equipped with Beeswing missiles. An example was pictured here ready for firing; the crew dispersed around the vehicle for safety, and the commander can just be seen at the top of the dune, from where he would fire the missiles. 101 FCC&R

KENYA

A batch of sixteen RHD 12V 101s was delivered to the Kenyan military through the Crown Agents in Kenya between July and September 1977. They had chassis numbers 957-00013A to 957-00028A and were probably used by the bodyguard of the Head of State, Jomo Kenyatta.

LUXEMBOURG

The Royal Luxembourg Army took delivery of sixty-one Land Rover 101s between 1975 and 1977, the first one being a demonstrator. Of the sixty further vehicles intended to meet the Luxembourg order, one (964-00395A) was burnt out just before delivery and an additional vehicle (964-00442A) was built to replace it. Two spare chassis frames were also supplied.

The Luxembourg vehicles were all 24V types and had chassis numbers:

964-00008A
964-00164A to 964-00203A
964-00383A to 964-00403A
964-00442A

Luxembourg was one of the first overseas users to show an interest in the 101, and took an early production LHD 24V GS model (964-00008A) for trials during 1975. This vehicle was equipped with a Nokken winch and was unique in that its front panel was never pierced for indicator lights. It was also equipped with a Scottorn powered-trailer drive, which differed from the Rubery Owen system that had been trialled by the British Army. (There are more details of this system in Chapter 3.)

A picture taken during the trials shows this first vehicle, numbered 3700 in Luxembourg, towing a Scottorn Bushmaster Mk2 powered trailer that has the registration plate EXC 384L, indicating it had once been used with development vehicle 961-00001A. However, the Royal Luxembourg Army decided not to take the powered trailer and ordered undriven one-tonne trailers from Scottorn instead. These were unusual in that they incorporated a number of 101 components – wheel rims, road springs and dampers. They are also thought to have been unique to the Luxembourg military.

The Luxembourg 101s were delivered in two batches, one in 1976 and one in 1977, to the army's main base at Herrenberg, just north of Diekirch. All were standard 964-series chassis with left-hand drive and 24V electrical systems and all were delivered in Matt Khaki paint (which has Land Rover code 275). Eight of the first batch had the Nokken winch. Most of these vehicles were intended as infantry transports, and so had longitudinal bench seats in the rear, intended to carry four solders on each side. However, five vehicles from this first batch were delivered with hard bodies that were built by Marshall's of Cambridge.

The second or 1977 batch was essentially the same as the first, although the GVW quoted on the chassis plate went up from 3,143kg to 3,652kg (6,915lb to 8,034lb), as on other 101s built at the

The Luxembourg trials vehicle, registered as 3700, is seen here with the Scottorn Bushmaster trailer that had originally been used behind EXC 384L, one of the pre-production 101s, and still wore that vehicle's registration plate. SCOTTORN

time. One vehicle in this batch was burnt out at the Solihull factory before delivery, and another one was supplied as its replacement. In this 1977 batch, nine vehicles were fitted with the Nokken winch.

All of the Luxembourg 101s had a tubular protection bar above the standard front bumper, plus mesh grilles over the front indicator, side and rear lights. Their front panels carried the standard mountings for pioneer tools, but the tools were never fitted. There were European inter-vehicle start sockets on the battery box covers, and the rear number-plate holders were relocated under the cross-member. Each vehicle carried a single jerry-can behind the front wheel on the driver's side, and each one had a map case inside the driver's door.

The five hard-body vehicles were on chassis numbered 964-00178A, 183A, 184A, 194A and 201A. They appear to have been delivered to Marshall's with standard GS bodies, and retained standard rear dampers when fitted with their new bodies. All five had the same design of body, which was quite different from the later Marshall's ambulance and signals types for the UK MoD. These box-bodies were separated from the cab, but communication between the two was possible through a small sliding window in each. The spare wheel was carried inside the cab behind the driver, and a removable panel gave access to it from outside. All five vehicles had a wire-mesh basket on the cab roof, which was used for camouflage netting and other items. Rear access was through two large side-opening doors and one fold-down flap, and a small ladder carried on one of the rear doors during transit could be hung on the fold-down section to aid access.

Three vehicles were configured for signals duties and the other two became ambulances. The three signals trucks, numbered 3755–3757 in the Luxembourg inventory, had a cable reel holder alongside the basket on the cab roof, and there was a clamp for a 12m (15ft) mast on the front panel below the passenger's side windscreen. The mast fitted into a socket on the front bumper, and was steadied by guy ropes when in use. Inside the body were an inward-facing bench seat on the right and a workbench along the front and the left-hand side. In the centre below the workbench were two extra batteries, charged from the auxiliary generating circuit of the 24V system, and a pair of Philips 40/60 radios stood on the front section of the workbench. Under the workbench on the left were two cabinets and an Eberspächer heater.

The signals trucks carried a steel penthouse frame on their roofs, and this folded out over the rear doors to create a covered work space. Above the cab, the front of the back body carried brackets for radio antennae, and the cables for these ran through tubular ducts in the cab sides and into the back body.

The ambulance bodies were on 964-00178A and 183A, and carried military numbers 3753 and 3754. Each had four stretcher positions, the upper pair served by a winding mechanism, which raised the

The Luxembourg colour scheme is clear in this picture of 3700, which remained in service as a runabout for some years after the others had been withdrawn. Interesting is that this vehicle never had flashing indicators fitted to the front panel. DARREN PARSONS

130 ■ THE 101 IN OVERSEAS SERVICE

This was the ceremonial handover of the first batch of 101s to the Luxembourg military. The taller of the two men in suits, who is pointing at a vehicle in one picture, is Land Rover sales manager **Jack Baines.** JACK BAINES/AUTHOR'S COLLECTION

stretchers into position. Alternatively, they could carry a mixture of seated and stretchered patients. The attendant had a small seat in the centre, above the gearbox cover, and the front wall had cabinets and holders for gas bottles.

The back body was heated by an Eberspächer unit located in a box bolted underneath it on the driver's side, and in front of this box was a single jerry-can holder. Strangely, the ambulance bodies had the same radio antenna brackets as the signals trucks, perhaps because it was easier not to modify the otherwise common body design. One of the ambulances (964-00178A) survives in enthusiast hands in Britain.

It appears that 964-00174A (3752) was fitted with a special body as a generating truck. One other vehicle (964-00400A, no. 3713) was used by the ordnance

ABOVE: **This Luxembourg 101 was brought back to the UK has been preserved by a member of the 101 Club and Register.** DARREN PARSONS

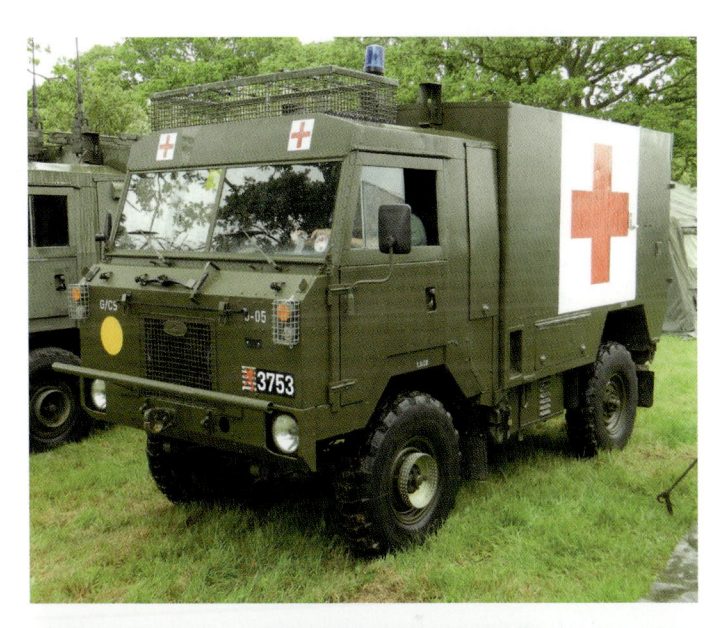

RIGHT: **The Luxembourg radio trucks and ambulances shared the same style of hard body, built by Marshall's.** AUTHOR

disposal unit, and 964-00384A (3746) was used for forward flight control duties. To this end, it was fitted with a second roll-over hoop at the rear that was used as a mounting for three radio antennae.

The Luxembourg Army made a number of interesting in-service modifications to their vehicles. For example, all the Nokken winches were stripped down to have a brake band fitted to the winch drive. This brake was actuated by a lever in the cab, located next to the driver's right leg. The system allowed loads to be held when the gearbox was in neutral, something that could not be done on the 101s used by the UK MoD.

Steering boxes were upgraded by replacing the ball bearings at the top and bottom of the worm drive with taper roller bearings that gave a higher degree of contact and smoother operation; proper oil seals were also fitted instead of the staked-in O-ring seal. The ends of the 24V shielded ignition leads were given conventional fittings so that standard 12V spark plugs could be used. Also, when exhaust systems wore out, they were replaced by stainless steel systems that the Luxembourg Army had made itself.

Disposal of the Luxembourg 101s began in 1992, when thirty-three were sold at auction. A further nine went in 1995, three more in 1998 and six more in 2000. Six more were donated to public departments, and at least three of these (including 3702 and 3729) were carefully converted to fire tenders for voluntary fire brigades, where they saw a further lease of life. This left only the trials vehicle, number 3700, on the strength, where it spent some time as a hack for the MT (Motor Transport) section at Herrenberg. At the time of writing, it was still there and was supposedly destined for the Luxembourg Army's historical collection.

OMAN

The Omanis took a total of forty-two LHD 12V 101s, beginning in 1976. Of these, thirty-seven were delivered to the Royal Guard at Muttrah in Muscat during 1976–77. The remaining five were initially delivered to Charles Kendall & Company in London during 1976. Now known as the Charles Kendall Group, this company has acted on behalf of the Oman Government since 1949, and its task was to oversee conversion work on the vehicles before delivery to Oman. All the vehicles were delivered in Matt Light Khaki paint (code number LM 171).

The thirty-seven direct-delivery vehicles all had front towing pintles and had chassis numbers:

959-00206A to 959-00218A
   (delivered September 1976)
959-00220A to 959-00230A
   (delivered April 1977)
959-00235A to 959-00247A
   (delivered September 1977)

The remaining five had numbers that filled in the gaps within this sequence:

959-00219A
959-00231A to 959-00234A

At least three of these five vehicles were fitted with special ambulance bodies by Pilcher-Greene in London, with an air-conditioning unit mounted above the cab. One of them – chassis number unknown – is now displayed at the Museum of the Sultan's Armed Forces in Muscat. A display board with it states that it entered service in 1976 and was withdrawn in 1989.

UGANDA

Just six 101s were delivered to the Ugandan military in June 1976, during the presidency of Idi Amin and

After withdrawal from military service, 3729 became a civil protection fire tender with the Commune de Bous. DARREN PARSONS

THE 101 IN OVERSEAS SERVICE ■ 133

The ambulances for Oman had a unique design of body, constructed in the UK by Pilcher–Greene. This one has been preserved at the Museum of the Sultan's Armed Forces in Muscat. 101FCC&R/LES ADAMS

shortly before Britain broke off diplomatic relations with the country. (Uganda had in fact broken off diplomatic relations with Britain five years earlier, in 1972.) The Ugandan 101s were RHD 12V models on chassis numbers 957-00001A to 957-00006A. They were delivered through the Crown Agents in Uganda and were equipped with Nokken winches, rear seats and hand throttles. No further details of these vehicles have yet come to light.

UNITED ARAB EMIRATES

A number of 101s were delivered to at least two destinations within the United Arab Emirates. The Land Rover despatch records show that thirty-two LHD 12V models were sent to the British Aircraft Corporation at Stevenage for onward transmission to Abu Dhabi. These were numbered 959-00103A to 959-00134A, had front towing pintles and were painted Matt Light Stone to suit desert use. However, only twenty-two appear to have been delivered, some fitted out as Swingfire anti-tank missile launchers. The remaining ten later entered British military service.

There were also three LHD 12V models (959-00381A to 959-00383A) that were delivered in March 1977 directly to the Armed Forces at Ras al Khaimah, which is the northernmost Emirate between Saudi Arabia and Oman. These were painted in Libyan Sand and had canvas tilts in stone colour, but nothing more is currently known about them.

UNKNOWN END-USERS

The end-users of some 101 One-Tonnes remain unknown to this day. The mystery vehicles make a total of forty-seven, and consist of two 956-series, six 959-series and thirty-nine 964-series. All the LHD vehicles (959- and 964-series) are known to have been delivered to the British Aircraft Corporation.

The two 956-series chassis are 956-00908A and 956-01305A.

The six 959-series chassis are 959-000135A to 139A. Their numbers follow on from those of vehicles sold to the UAE, and this could imply that they were a cancelled UAE order. Possibly, and more likely, is that they include the Swingfire vehicles delivered to the Egyptians.

The thirty-nine 964-series vehicles are:

964-00038A
964-00072A
964-00158A to 964-00167A
964-00204A to 964-00230A

Their 24V LHD specification makes it likely that they became Rapier vehicles for an unidentified overseas military force, but so far no evidence has been discovered.

There is an unconfirmed story that five 101s found their way into Iranian military use in 1976–77, and these vehicles (if they existed) could have been drawn from the batches listed above.

UNSUCCESSFUL OVERSEAS TRIALS

Solihull provided examples of the 101 to several overseas countries for military trials, but not every trial produced an order. The 1971 Malayan trials (covered in Chapter 2) and the 1973 Canadian trials (covered in Chapter 3) were examples of trials that did not have a successful outcome; the Australian trials in 1969 (discussed earlier in this chapter) can also be added to this list. One of the prototypes (see Chapter 2) was originally planned as a demonstrator for the South African military, but the plan was aborted. One of the pre-production vehicles went out to Leykor in South Africa some four or five years later, perhaps in fulfilment of the original plan (see Chapter 3). However, no order followed.

The Malayans would have placed an order if Land Rover had been able to give them a firm delivery date. However, this was simply not possible because there was at that stage no final production specification. Nevertheless, the Malayans did not give up. As late as July 1977, there seems to have been a plan to build a 957-series (12V, RHD, export) vehicle for them. However, this was never built, probably because British Leyland did not want to prolong production of the 101.

Rumours of a demonstration to the Belgian military seem to have no foundation in fact. They probably result from confusion with a project to offer 110 Forward Controls to that country in 1969.

CHAPTER EIGHT

AFTERLIFE

Once their military service was over, the 101 Forward Controls gradually filtered down to civilian ownership by way of specialist dealers and auctions. They proved very much in demand, perhaps surprisingly so for vehicles that are quite Spartan in nature and not particularly easy or comfortable to live with. In Britain, disposals went on through much of the 1990s, and it was no surprise that an owners' club was formed at the end of the 1980s.

Membership of this club, the 101 Forward Control Club and Register, is absolutely indispensable to anyone who owns or is planning to own a 101 One-Tonne. Its members have carried out a great deal of research over the years, not only into the history of the vehicles but also into the practicalities of ownership. The club has arranged for the remanufacture of some spares, and has looked into alternatives for some items that simply cannot be replaced. Understandably, the club's remanufactured items are available only to members, and only members receive its magazine, which is called *Six Stud*. The best place to start is with the club's web site, at *www.101club.org*.

Another use for a 101: when the Amarula drink was first imported into the UK, this 101 was fitted out as a promotional vehicle. AUTHOR

This signals-bodied vehicle was used for a time by the safety team at the Channel Tunnel Rail Link. 101FCC&R

Nevertheless, the 101's post-production history has not only been about enthusiast ownership. The basic design was further developed by MSA in Spain, who sold it on both the civilian and military markets during the 1980s. Several 101s were bought by tour companies in southern Africa, where they saw further use as safari vehicles – sometimes quite extensively modified for the purpose. Later still, a quantity of 101s were used to create some elaborate film props – and many were sold off in film-prop condition.

THE SPANISH CONNECTION

Since the late 1950s, the Rover Company (and Land Rover Ltd after 1978) had enjoyed a close working relationship with Metalurgica de Santa Ana, or MSA, who used the trading name of Santana. With head offices in Madrid and a factory at Linares in southern Spain, this company built Land Rovers for local sale from CKD kits sent out from Solihull. Part of the agreement with the Rover Company was that the proportion of Spanish-sourced components should gradually increase over the years until the whole vehicle was actually made in Spain. That stage was reached in 1967, by which time Santana was also exporting its Land Rovers to several countries in Africa, the Middle East and South and Central America, where Spain had established trading links.

Even though the Santana manufacturing operation was now self-sufficient, and even though the Spanish company was also developing its own variations on Land Rover designs, it retained close engineering links with Solihull. Indeed, Santana designs were used on UK-built Land Rovers as late as the 1980s, when V8-powered One-Ten and Ninety models were supplied with the Spanish-designed LT85 five-speed gearbox, and the winding door windows introduced in 1984 were of Spanish design, too.

So it was not very surprising that the design of the 101 One Tonne was handed over to MSA after production ended at Solihull in 1978. Strangely, almost no evidence has been discovered so far to shed light on how the handover was carried out. Design drawings must have been sent to Spain, but whether any production tooling went to MSA remains unclear.

The Spanish plan was to turn the 101 into a replacement for their existing Forward Control model, which was called the Santana 1300 and was based on the Land Rover Series IIA 109 Forward control. They re-worked the design, replacing the V8 engine with their own 3.4-litre OHV 6-cylinder petrol and diesel engines (which were actually 6-cylinder derivatives of Land Rover's own 2.25-litre 4-cylinders). They added to these their own gearbox and transfer gearbox. For the road springs, they used standard semi-elliptics, rather than the taper-leaf type used on the 101.

AFTERLIFE ■ 137

The 101 parentage of the Santana 2000 is most recognizable in this shot of the chassis. The engine was Santana's own 6-cylinder. MSA

One example of the Santana 2000 was brought to Land Rover's Solihull headquarters for evaluation. It was pictured in the UK prior to sale through a Land Rover dealer.
ROGER CONWAY

138 ■ AFTERLIFE

Santana had found a market for double-cab models with their earlier 1300 Forward Control, and built a number of 2000s with a similar configuration. MSA

Further work produced a new and larger cab and a redesigned back body as well. With twin headlamps and a slatted grille, the new model was barely recognizable as a descendant of the 101 One-Tonne, or indeed as a Land Rover of any kind. It was rated for a 2,000kg (4,400lb) payload (twice that of the UK-built 101), and was introduced to the Spanish market during 1979 as the Santana 2000. It appears to have been the first Land Rover-derived model from MSA not to have worn Land Rover badges. Just one example came to the UK in 1979 for evaluation at Solihull. Originally registered as BOM 99V, it was later sold off through a UK Land Rover dealership and went to Ireland, where it was re-registered as GJI 1045. It spent its days as a milkman's delivery truck and by the mid-1990s was derelict in Londonderry.

The Santana 2000 was available with a variety of body configurations, including a four-door crew cab with pick-up rear body. Many were used by local authorities, and some were configured as fire tenders. Santana also offered the vehicle on the military market with the name of S-2000. Sales brochures pictured a GS soft-top variant that was probably intended as a troop carrier; a hard-body variant possibly intended for signals use; a gun portee with what looks like a Hispano-Suiza or Oerlikon 20mm anti-aircraft weapon; and a lighting truck that incorporated a generator in the rear. Whether all these versions were actually built in quantity and, if so, who bought them, is simply not known.

One way or another, the Santana 2000 was not a big seller. Figures from Santana show that just 923 were built before production finished in 1990 – an average of around eighty-four vehicles a year. All of them appear to have had left-hand drive. Early chassis numbers carried the prefix E689, but it is likely that later vehicles carried standardized VINs. Coincidentally, it was in 1990 that Land Rover Ltd disposed of its stake in MSA, claiming that the two companies now built such different vehicles that there was no longer any point in co-operation.

THE *JUDGE DREDD* 101S

During 1994, a total of thirty-two very special vehicles were built on ex-military 101 chassis for the science-fiction film *Judge Dredd*, starring Sylvester Stallone. The 101 chassis was chosen partly for its height off the ground and partly for its ready availability, as the British armed services were then beginning to withdraw their vehicles in large numbers.

The original design was inspired by some futuristic sketches that had been drawn up by David Woodhouse in the Land Rover Design Studio. The film's

AFTERLIFE ■ 139

The Santana 2000 also became a military vehicle, although some of the examples in these pictures may have remained unique demonstrators. MSA

140 ■ AFTERLIFE

This was a concept sketch for the *Judge Dredd* film vehicles, prepared at Land Rover. LAND ROVER

This was Land Rover's own publicity picture of a *Judge Dredd* 101. Note how the wheels are made to look wider by GRP facings. LAND ROVER

director saw them, liked them and asked for them to be refined to suit the film he was planning. The idea was that the vehicles would represent the armoured transport necessary in twenty-second-century New York.

Once a design had been agreed, Land Rover built a prototype body at Canley from GRP and mounted it on a 101 chassis (964-00462A, 75 GJ 20). This vehicle was built as a 'taxi' (painted yellow to suggest New York's famous Yellow Cabs) with a fully trimmed interior and was the only one of the *Judge Dredd* 101s to have one. All the later examples were simply hollow shells.

A contract to build a further thirty-one film vehicles was arranged with Dunsfold Land Rovers (now DLR). Dunsfold's Philip Bashall scoured the auctions and small ads for suitable vehicles, and when the Land Rover community heard about this sudden interest in 101s, a rumour began to circulate that the MoD was in fact buying the vehicles back through a 'front' company because the RB44s it had bought to replace its 101s were so unreliable.

AFTERLIFE 141

RIGHT: **The original one-piece GRP body moulding is seen here on arrival at Dunsfold Land Rovers, prior to mounting (with difficulty) on a 101 chassis.** DUNSFOLD DLR

BELOW: **All three types of *Judge Dredd* vehicle were lined up for this photograph, taken within the grounds of Land Rover's Solihull factory.** LAND ROVER

The contract to make the GRP bodies went initially to Wood & Pickett, the company that had been custom bodywork specialists. This company was then under new ownership and was based in the West Country. A mould was made from the prototype, and about four early bodies were produced. However, when Dunsfold came to fit the bodies to the stripped 101 chassis, it became clear that this was simply not possible. The upshot was that the body contract was re-allocated to Futura in Coventry, who redesigned the body to be made in two halves, which then bolted together.

Most of these bodies were painted yellow to represent taxis; a few were red or silver, to represent service and amenity vehicles. The exact figures are unknown, but the probability is that there were twenty yellow ones, and six each of the red and silver vehicles. After the filming had been completed at Pinewood Studios, the vehicles were returned to Land Rover for disposal. Some were sold off for

Scenes from the *Judge Dredd* film. The setting is New York in the future.
BUENA VISTA PICTURES

promotional use, while others were rebuilt to their original military condition by Dunsfold Land Rovers and then re-sold. The fully trimmed vehicle now belongs to the vehicle collection at the Heritage Motor Centre in Gaydon.

OWNING A 101 TODAY

Anybody who is contemplating ownership of a 101 today will want to know what the vehicle is like to drive. Usually, driving impressions of older vehicles are fairly readily available through magazines that were published when the vehicles were new, but that is not the case with the 101. As it was a military-only vehicle, very few driving impressions appeared in print until after examples began to reach the civilian market in quantity during the 1990s. When they did, they were usually in the specialist off-road magazines and not in the mainstream publications.

One early release from military service was tested by Brian Hartley in *Off Road & 4 Wheel Driver* magazine for October 1985, who had this to say about it:

> *The 101 is as beautifully grotesque as only a fully fledged Army vehicle can ever hope to be. Yet, in relation to the tiny numbers ever made, it has a 4WD fan club (myself included) out of all proportion to its importance. Dedicated followers of 4WD fashion could point to its slab sides, bluff front and torture chamber cab somewhat derisively. I will point out massive ground clearance, short wheelbase, stupendous approach and departure angles, visibility second to none and a power house of an engine. In short, an off-roader's dream come true!*

ABOVE: **Sylvester Stallone played Judge Dredd in the film.** BUENA VISTA PICTURES

RIGHT: **After use in the film, some of the *Judge Dredd* vehicles were rebuilt as standard 101s. Others kept their special bodies and were used for publicity purposes. This one carries the name of Futura, who built the two-piece GRP bodies.** AUTHOR

The cab...appears to have been designed to inflict maximum pain to the greatest number of people.... The steering wheel and its positioning is pure 'truck' and none the worse for that. The gearstick sprouts up from the engine cover close to your left hand whilst instruments and minor controls are all sensibly placed. The biggest drawbacks are in having to put a leg either side of the steering column and not being able to move the old plates of meat rearwards, a fault shared with many forward-control layouts.

The biggest advantage is vision. Until you have driven one you can never know what joy it is to be able to see the ground in front of you and not obscured by some hulking great bonnet. This means that you can place the vehicle far more accurately when off-roading and, when cresting hills, you can see over the top, even before the vehicle rumbles over the edge. Conversely, coming down steep hills you quite often feel as if you are about to be splattered on the windscreen.... The ponderous gearchange needs 'mansize' movements to engage the gear of your choice, brakes are excellent going forward and abysmal, verging on the non-existent, in reverse. The steering, however, is very light and precise.

Like all load carriers the 101 tends to hop a bit when unladen, but the ride is quite comfortable.... The seating position, directly over the front wheels, tends to exaggerate any vertical movement, particularly on side slopes, but though it feels quite hairy, the centre of gravity is low down...and the 101 is surprisingly stable on side slopes; even on three wheels which is common with stiff springing.

The V8 is ideally suited for the vehicle, its power allowing you to trickle over obstacles, with plenty of grunt to give a 'squirt' at the right moment, to lift the vehicle's statuesque bulk over or through a tricky section.

The side mounted winch is a highly unusual but practical piece of kit...the winch is well out of harm's way, easy to reach for service or repair and ideal for feeding the winch rope either forward or rearward over its guides.

Brian's comments were spot-on. The 101 is indeed a superb vehicle to drive off-road, although its size can make it rather a handful in town traffic. Fuel consumption of around 13–14mpg (21.8–20.2ltr/100km) is a major deterrent for many would-be owners, and there is no doubt that cab access is not for the less than agile. Cab space is cramped for tall drivers, too; then there is the issue of that gearchange, which takes some acclimatization, but is in fact not a problem once the driver has learned to trust it. Not everybody finds it easy to like the flat, bus-like angle of the steering wheel, although it does give plenty of leverage for the rather heavy steering.

However, driving a standard 101 is a piece of cake compared to driving one of the modified 101s that were used in the *Judge Dredd* film. The present author was given the chance to drive the only vehicle with a complete interior over some sections of the Land Rover experience off-road course at Solihull – an experience recorded in *Land Rover Enthusiast* magazine for August 2005. Quite apart from the reduced forward visibility through the slit windows, the driving position was the next best thing to hell on earth for somebody 6ft 2in (188cm) tall – and getting into and out of the vehicle required both uncomfortable contortions and considerable assistance!

BUYING A 101

Why buy a 101? There are many reasons why people want to own one of these very special Land Rovers, and not the least of them is that they are relatively uncommon. The 101 also has an undeniably rugged image that appeals to some people. Its large load area makes it extremely versatile, too. But one thing is certain – a 101 is not ideal as everyday family transport. It was never designed for that purpose.

Many people want to buy a 101 because they have an interest in military vehicles, and a percentage of those buyers will try to restore an example to the way it would have been when in military service. Others prefer to restore it to the way it would have looked when it left the assembly lines at Rover –

AFTERLIFE 145

This signals-bodied vehicle was used by Devon Renewable Energies for a time and the large tank under the body was presumably for LPG. It was pictured at the Great Dorset Steam Fair in the late 1990s.
AUTHOR

which was quite a different thing. Some of the special-bodied vehicles, such as signals-body and ambulance types, present their own special challenges, and there are buyers who enjoy those challenges in their own right.

Of course, many buyers have seen in the signals and ambulance-bodied versions of the 101 the perfect basis of a camper, and there is no doubt that the hard-body vehicles provide an ideal enclosed space for that job. Others have built their own camper bodies onto standard GS trucks. One point that comes from this is that such conversions ensure that the numbers of really good vehicles in standard condition will gradually diminish over time – which may or may not increase prices and add to their desirability.

Running Costs

The first thing to consider when planning to buy a 101 is the running costs. The major expenditure will be on petrol: a 101 with its original 3.5-litre petrol engine and four-speed gearbox generally consumes fuel at the rate of 14mpg (20.2ltr/100km) or so. That makes it expensive for long journeys in particular, and is one reason why some owners have converted their vehicles to run on LPG (which in Britain at the time of writing cost about half as much to buy as petrol).

However, LPG conversions in themselves can be expensive, and it takes a lot of miles before that initial cost is saved in fuel expenditure. Some owners have even converted to diesel engines, typically Land Rover's own 2.5-litre 4-cylinder Tdi types, but once again the vehicle has to be used for a lot of miles before

the cost of the conversion is recouped through fuel savings.

Servicing costs need not be a worry, though. The 101 is in many ways ideal for DIY maintenance, because it was designed to be simple to work on in the field. For working underneath the vehicle, the 101's great height is an obvious advantage.

Engine access is slightly tricky, because the 3.5-litre V8 is somewhat buried between the front seats. A metal panel has to be removed first, and even then access is certainly not easy. Service items such as filters are readily available and are not unreasonably priced. Perhaps most important is to remember that the V8 engine needs much more frequent oil changes than most modern engines, and 6,000 miles (9,600km) should be the absolute limit between changes. Transmission oil should also be changed every 24,000 miles (38,400km), or earlier if the vehicle has been wading because of the risk of water contamination.

Inspection

It will be pretty obvious from a casual glance whether a 101 offered for sale has been abused as a heavy duty off-roader, cherished as an original-condition example, modified to suit the owner's wishes and so on. Some vehicles will have been modified in small ways that are designed to add to comfort and convenience, and many of these modifications are reversible. So, for example, adding later Land Rover front seats with head restraints can be considered a desirable safety and comfort improvement.

Perhaps most important on an initial examination of any vehicle is to assess how much work needs to be done to get the vehicle into the condition required. Getting a poor runner into good running order will probably be less challenging and less expensive than trying to turn a battered and well-used example into something more presentable. Missing items, such as the Nokken winch, will be both difficult and expensive to replace.

Bodywork

The 101's bodywork is the usual Land Rover combination of aluminium alloy panels and steel reinforcement. The alloy panels are easily damaged but, although gashes can be unsightly, owners tend to ignore minor scuffs and dents. Steel items can rust, and rot in the lower sections of the detachable door tops is common. The side panels behind the cab doors are also made of steel and can rust through from behind.

The condition of the canvas hood is important, because replacements are expensive. However, there is no point in worrying about whether it will keep out draughts and leaks, because it simply will not.

Chassis

Inspecting the chassis is straightforward, thanks to that high ground clearance. It is certainly rugged, but even though it was undersealed from new, military-style, it is certainly not impervious to rust. The rear cross-member quite commonly rusts out. Rainwater is channelled onto it by the two brackets that carry rubber-faced over-riders. Still at the rear, the spring hanger brackets can suffer – and it is not unknown for the front ones to rust in bad cases, too. Another common rust area is the outrigger in front of the fuel tank.

In the Cab

The cab was always designed to be functional rather than attractive, so there will be wires visible under the dashboard. The brake servo is exposed under the driver's side (where it is also very accessible), and the fluid reservoirs for brake and clutch sit alongside the instrument panel. Even the washer bottle is exposed, down in the passenger's side footwell. There were holes for three auxiliary instruments in the dashboard, and at least one of them is likely to be empty; there was no such thing as a blanking plug. One item unfamiliar to those new to military vehicles will be the main lighting switch: one position turns off all the lights except the headlights (which could be fitted with infra-red equipment) and another gives only the convoy light (which was under the rear of the vehicle and shone on the white-painted back face of the differential, so that following drivers could see it).

The condition of the seats is worth checking, and it is also worth looking at the trunking that connects the heater box to the channel running on top of the

This former Australian Rapier tractor was converted into a very professional-looking camper. It was seen here during the Grand Parade at the Land Rover 50th Anniversary celebrations in Cooma, New South Wales, in 1998.
AUTHOR

engine cover that distributes warm air into the back body. Leaks and poor connections will not add to comfort for the occupants.

In the Back Body
Some vehicles have inward-facing seats in the rear, while others do not. The seats were not designed to take safety-belts. It is worth checking that the rear engine cover (at the front of the body, behind the cab) is not only present but properly secured. Otherwise, the condition of the floor and the wheelboxes will give a good idea of how the vehicle has been treated.

On GS vehicles, the hood sticks should all be present and in good condition. The spare wheel should be secured by a central clamp behind the driver's seat, and not loose in the back. The tailgate should have a chain on each side, too. The box-body vehicles had various types of interior lining, and this should be present and undamaged. On ambulances, the most common box-body type, it is a light green Formica material designed to be wiped down easily, and should not have cracks or missing sections.

The Electrical System
The electrical system is very simple by modern standards, but age means that the wiring loom may be past its best. If it has been hacked about so that additional equipment can be fitted, be particularly wary. Replacement looms were available at the time of writing. Corrosion is quite common around the electrical terminals and in the fuse box, but can usually be cleaned up successfully without much effort.

Some 101s were built with 24V electrical systems. These are wholly reliable, but it is worth remembering that parts as simple as spark plugs cannot be obtained from normal parts outlets. The distributor, too, is a double-points type that can be tricky to work on. Many owners have converted to 12V systems, but beware of half-hearted conversions that have been done on the cheap.

The Engine
When healthy, the 3.5-litre V8 engine is delightfully smooth and powerful-sounding. However, it does need regular oil changes, and neglect in this department often leads to premature wear of the camshaft or to failure of the hydraulic tappets that take up the clearances in the valve train. Even engines in good condition may rattle briefly when first started from cold, as the oil takes time to circulate and pump the tappets up. A persistent rattle points to problems.

The oil pressure in these engines is usually quite low, but if the oil light on the dash flickers at idle when the engine is warm, that pressure is too low. There is in fact a safety cut-off switch which will prevent fuel from reaching the engine if the oil pressure is too low. Serious neglect of the oil level may lead to main bearing wear, which is apparent as a rumble from the engine when on the move.

This example started out as a standard GS truck, and was later fitted with a home-made box-body for use as a camper AUTHOR

An engine that overheats or loses coolant may be a warning of head-gasket failure. White exhaust smoke will confirm that this is the problem, and uneven running may be another symptom. One way or another, it is important to get to the bottom of overheating or coolant loss problems, because they can cause all kinds of mayhem in this all-aluminium alloy engine.

Other problems may be associated with the exhaust manifolds, which can crack and create a distinctive chuffing noise. The end seals of the valley gasket can leak, allowing oil to gather on the top of the engine, but fitting the improved gasket introduced for later versions of the engine usually cures this relatively cheaply.

The Transmission

The two-speed transfer gearbox is integral with the LT95 four-speed primary gearbox. It is not unknown for an internal seal to fail and for the transfer box to dump its oil into the main box, so starving the high-ratio gears of lubrication. Any unusual noise, particularly grinding noises, is a sign that something is wrong. However, the LT95 was not known for being a quiet gearbox, and a fair amount of whine at speed is normal.

The gear lever has a long throw and is somewhat awkwardly placed, but gear selection is clean as long as changes are not rushed. Any gearbox that jumps out of gear on the over-run has a problem. The synchromesh between first and second gears was always baulky when cold, and some owners have used engine oil (thinner than the recommended type) to alleviate this. The transfer gearbox selector depends on a cable, which can seize. Wear in the transfer gears may cause the ratio to jump out here, too.

Propshaft rumble is common on 101s, and results mainly from the very steep angles at which the shafts operate. However, persistent cases deserve further investigation; a likely cause is worn UJs in the propshaft. Twisting them backwards and forwards to check for excessive movement is the best way of ruling this one out. Worn splines can be detected by pushing the propshaft up and down in the centre. Split propshaft gaiters are another result of those steep angles.

Like all Land Rovers, the front axle on the 101 uses a chromed ball swivel at each end. Over time, this can become pitted, so allowing lubricating grease to leak out through the wiper-type seals. As new 101 axles have not been available for some time, professional repair is the best solution. It may be expensive, especially if the ball swivels have to be re-chromed.

Suspension, Steering and Brakes

The ride in a 101 is surprisingly smooth for what is fundamentally a small truck, and that has a lot to do with the taper-leaf springs. However, an unladen vehicle can feel rather stiffly sprung at the rear. The springs themselves are usually trouble-free.

Many people find the steering heavy at low speeds, and it is almost impossible to turn the steering wheel at all unless the vehicle is moving. In fact, trying to do so when the vehicle is stationary can damage steering components. Steering boxes are actually a weak point, not least because water can get into them; any vertical movement of the steering wheel means that the steering box is on the way out. If there is a rough, grating feel as the steering wheel is turned, the bearing at the top of the steering column may need replacement. All 101s wander a little at speed, but excessively vague steering may point to a damaged steering box, or perhaps to worn bushes on the front anti-roll bar.

As for the brakes, the all-drum system is easy to work on but it does have a couple of weak points. One is the load-apportioning valve, which is designed to prevent premature lock-up of the rear wheels when the vehicle is travelling unladen. A skittish back end under braking suggests there is trouble here: the valve can seize, and some owners remove it altogether rather than replacing it – a modification that is both dangerous and illegal. The second weak point is the brake servo, which has a Bakelite piston that breaks up. Replacements have been next to impossible to find for a long time, but it is possible to use the readily available servo from a Land Rover Defender 90, with a little ingenuity.

Tyres were originally Michelin XZL types, but these have not been available new for some time. The 9.00 × 16 size is uncommon now, and can be difficult to find. Very important is to check the speed rating of tyres in this size: some are agricultural types with a maximum speed rating of just 30mph (48km/h), and a 101 should be capable of twice that speed (although rarely more).

APPENDIX

POWERED-AXLE TRAILERS

The idea of a 6×6 power train that incorporated driven wheels on the gun carriage was very much part of MoD thinking by the mid-1960s, and was already firmly in place by the time the two 110in gun tractor prototypes were built. As a result, it had a major influence on the early stages of the new Military Forward Control project that would deliver the 101 One-Tonne. However, for a variety of reasons, the system was not carried over to production vehicles.

One result has been a great deal of misunderstanding about the role of the powered-axle trailer in the One-Tonne story. So what follows is an attempt to put the record straight, compiled with valuable input from Paul Hazell, whose article in *Windscreen* magazine for Spring 2011 was probably the first

Now owned by the Heritage Motor Centre at Gaydon, this 101 with powered-axle trailer was pictured during a Land Rover press event at Eastnor Castle. LAND ROVER

proper attempt to research the trailer story. Also invaluable as a source has been the chapter written by George Mackie for Ken and Julie Slavin's book, *Land Rover, The Unbeatable 4×4*. George was the head of the Land Rover Special Projects' Department, which was closely involved with the development of the powered-axle trailers.

IN THE BEGINNING

Land Rover became involved in experiments with a powered-axle trailer during the early 1960s. The main reason was that trailers tended to cause the towing Land Rover to lose traction in rough terrain. As George Mackie commented:

> The loss of traction always seemed out of all proportion with the fact that there was at least still four-wheel drive in a six-wheel unit.
>
> So the logical next step was to see if it was possible to power the trailer axle so that it contributed to the drive instead of working against it. The initial work was entrusted to Stephen Savage, then a Land Rover apprentice. However, the project had not gone far before George Mackie met Major Kitchen, the sales manager of trailer manufacturer Scottorn Ltd, at the Commercial Motor Show in autumn 1960. One discussion led to another, and the upshot was that Scottorn agreed to take over further development of the powered-axle trailer. The work would be done in conjunction with Land Rover, would use Land Rover parts and, if a viable product resulted, would gain Land Rover approval. This was seen at the time as a valuable aid to sales.

THE SCOTTORN DESIGN

Within a couple of years, Scottorn had a design ready. The key figure in its design, recalled George Mackie, was Rex Sewell, an ex-Chobham man, retired and a specialist in trailers, who was taken on by Scottorn as a consultant to do the final design work on the trailer drive.

The Scottorn system provided a simple drive-coupling for the trailer at the rear of the Land Rover. Power reached this coupling through a propshaft driven from the centre power take-off on the transfer box, and that propshaft drove an auxiliary gearbox mounted just above the rear cross-member. In the beginning, this auxiliary gearbox was a modified Land Rover transfer box, but it appears that a dedicated unit was designed later.

The trailer was attached to a standard tow hitch in the usual way. Running above the drawbar was a propshaft attached to the trailer's differential, and this propshaft was then coupled to the power drive on the back of the Land Rover. The trailer's axle was another off-the-shelf Land Rover component, in this case the rear axle from a Series IIA 109in model. The trailer drive allowed operating angles of up to 60 degrees between Land Rover and trailer in all planes.

This system allowed the driver to select two-wheel, four-wheel or six-wheel drive to suit the terrain. To prevent different ratios being selected in the two transfer gearboxes, they were interconnected by a linkage to the selector lever in the cab. The Scottorn design was filed as a patent in October 1962,

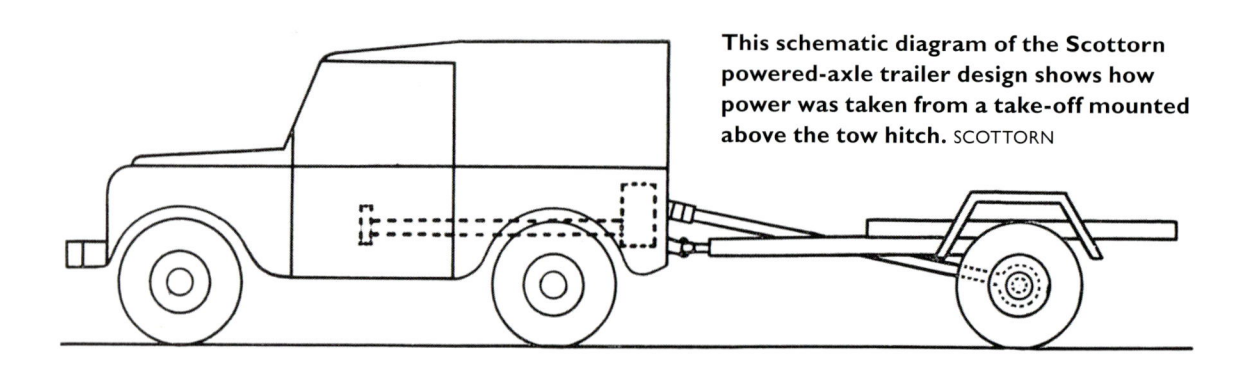

This schematic diagram of the Scottorn powered-axle trailer design shows how power was taken from a take-off mounted above the tow hitch. SCOTTORN

together with a simplified hand-drawn schematic diagram showing the layout of the system. Not long afterwards, the Scottorn powered-axle trailer was made available commercially as the Bushmaster.

The Bushmaster had a welded chassis with the drawbar bolted to it, and its Land Rover 109 axle came with freewheeling hubs to eliminate frictional losses when the trailer drive was disconnected. The springs were standard Land Rover types with Aeon rubber assisters. The trailer could be fitted with a variety of different bodies, and indeed one version of it found its way into British military service in the later 1960s.

FVRDE at Chobham examined a Bushmaster trailer in late 1964 or early 1965. As Paul Hazell has explained:

> It seems the trailer was thought of as an ideal solution to the problem of a tactical [airfield] crash tender, by virtue of it being able to double the payload of a Land Rover with little loss of mobility in rough terrain, while making the vehicle divisible for airlifting.

So during 1966, at least two and possibly three Rover Mk9s (military Series IIA 109 models) were modified with a trailer-drive transmission, and the same number of cargo-bodied Bushmaster trailers were tested with them. Although the primary objective was to test the feasibility of the crash tender idea for the RAF, FVRDE was also more generally interested in the concept of a powered-axle trailer. By late 1967, the concept had been proved sufficiently for fifteen Land Rover 109s to be prepared as towing vehicles; over the next two years a prototype was assembled with Pyrene fire-fighting equipment on the towing vehicle and a 183-gallon (832ltr) water tank on the trailer.

Tests completed, the vehicle-and-trailer combination was signed off for service and all fifteen examples had entered service by 1972. They were known as TAC-T types, the acronym deriving somehow from the full name of Truck, Fire Fighting, Crash, Tactical ¾-ton Rover 11 with Trailer, Tanker, Driven Axle, Water ¾-ton, 2W, 200 gallon, Scottorn. In service, the TAC-Ts were used to provide fire and crash cover for Harrier jet temporary airfields, and also for fire and crash cover at some permanent airfields.

THE FVRDE/RUBERY OWEN SYSTEM

However, even though the Bushmaster trailer had been accepted for RAF service in a specialized role, FVRDE were not sure that it would withstand life elsewhere in the armed services. In particular, they wanted it to be capable of being jack-knifed and rolled through 360 degrees; without that second requirement, if a trailer turned over it could theoretically take the towing vehicle with it. Quoting George Mackie again:

> Chobham [i.e. FVRDE] decided that they would have to design something [themselves]. In due course they did: they produced a very elegant design, which catered for 360 degree angularity, but demanded a specially adapted vehicle for attachment purposes, and a special trailer, neither of which were compatible with other units already in use.

The new coupling was developed by FVRDE and its manufacture and further development were handed over to Rubery Owen Ltd of Darlaston in Staffordshire, who were the British Army's favoured supplier of trailers. Their powered-axle trailer was first shown at the 1966 SMMT–FVRDE exhibition at Chertsey, when it was still under development. Here, it formed Exhibit number 21A, alongside Exhibit number 21, a 110 Gun Tractor fitted with the new concentric drive coupling. The Scottorn trailer was there too, as part of Exhibit number 15.

Like the Scottorn trailer, the Rubery Owen design had steel-welded construction with semi-elliptic leaf springs and rubber spring assisters (in this case by Avon) on a rigid axle with differential. The 1966 Chertsey catalogue explains that:

> The tow eye and power driven coupling are interchangeable to suit the role. When the tow eye and hook are fitted the braking arrangement is by an over-run mechanism but if [the] power driven coupling is fitted the brake operation is vacuum/mechanical. The power driven trailer role when coupled to a Land Rover four-wheel drive vehicle

gives a six-wheel driven train with considerably higher mobility over soft terrain particularly if gradients are to be climbed.

Where the Scottorn system kept the tow hitch and the power coupling separate, the Rubery Owen system worked by combining the trailer hitch and the rotating drive coupling into a single unit. This unit had to be aligned with the receiver in the towing vehicle's rear cross-member, and was then pushed into it. Deep within the receiver was a rotating unit with splines that engaged with grooves inside the trailer hitch. The splined receiver unit was spring-loaded, which allowed the splines to engage automatically as soon as the propshaft began to rotate.

This arrangement provided the drive to the propshaft on the trailer, and it was supplemented by a system that kept the tractor and trailer coupled together. Behind the trailer hitch was a cast flange, and around the receiver socket on the towing vehicle was a second flange. When trailer hitch was inserted into socket, these two flanges came together. To keep them together, a length of special chain was wrapped around them and then tightened by a device like an over-sized wing-nut. As long as the chain stayed in place, the coupling was secure. Drivers were advised to stop after 500yd (460m) to double-check that it really was.

The driveshaft on the trailer ran from the hitch to the axle differential. At the front end, it had a pair of universal joints that allowed for changes in the angle between the vehicle and the trailer, and also allowed for some play in the tow-bar yoke. The yoke allowed pitch and yaw of up to 60 degrees and could theoretically handle a full 360 degrees of roll – which would have seen the trailer turn right over and then right itself again.

The FVRDE/Rubery Owen design for a powered-trailer drive depended on this coupling (*above*) on the nose of the driven trailer. It had to be carefully aligned with the receiver hitch on the towing vehicle (*left*), which was no easy task.
IMAGE ABOVE: AUTHOR;
IMAGE LEFT: ROVER CO LTD

POWERED-AXLE TRAILERS ■ 153

The receiver hitch was built into the rear cross-member of the towing vehicle. Also visible here is the vacuum tank for the trailer brakes. AUTHOR

If the towing vehicle was to be used with an unpowered trailer, a standard NATO tow-hitch could be secured into the opening of the coupling on the rear cross-member. When a powered trailer was being towed, the NATO hitch could be stowed in a 'keep', which consisted of a pair of elliptical brackets on the left-hand rear body panel. All of the first seven prototypes of the 101 One-Tonne had this, although it was not fitted to later vehicles with the trailer drive.

On the original 6-cylinder 101 prototypes, power came from a 'bottom' PTO on the gearbox because the usual centre PTO was dedicated to driving the

A standard NATO hitch could be fitted in place of the powered coupling. It was secured by a chain around the mating flanges, in exactly the same way as the powered type. AUTHOR

LEFT: **The prototype trailers were also built by Rubery Owen. This is the identifying plate on one of the pair that crossed the Sahara behind 101 One-Tonne models in 1974.** AUTHOR

OPPOSITE PAGE: **This was how everything looked when the powered trailer was hitched to a 101 One-Tonne Land Rover. The complicated and expensive design of the trailer hitch and drawbar is obvious.** LAND ROVER

front-mounted winch. When the V8 engine entered the specification, a similar arrangement was used, even though the winch was a different type and was now mounted at the side of the vehicle.

As for braking, a vacuum linkage had to be attached between trailer and towing vehicle, where the connection was mounted on the rear cross-member. Just behind that cross-member was a vacuum tank, mounted across the rear of the 101's chassis. This was linked to a vacuum reaction valve (made by Clayton Dewandre) that was fitted to the chassis in the hydraulic line close to the brake apportioning valve. The vacuum in the reservoir tank was maintained by the engine, and normally held the trailer brakes in the off position. However, as the brake pedal was pressed and fluid was moved under pressure through the hydraulic lines and the vacuum valve, the valve released its vacuum and so applied the trailer brakes.

The MoD decided at an early stage that they wanted to use the Rubery Owen powered-trailer system on the 101 in preference to the Scottorn type. So the 101 prototypes had it from the beginning. About ten trailers were made in all, with minor variations in the hitch arrangements. The first examples were built with an ENV axle that had five-stud hubs, but later ones switched to the Salisbury axle and six-stud hubs that had been settled as standard for the 101 itself in 1970.

Scottorn, meanwhile, realizing that they were about to miss out on a potentially lucrative contract, came up with what they called a Bushmaster Mk2, which was based on the Mk1 and did not use the FVRDE-designed coupling, but had a revised transfer box in the towing vehicle and a repositioned trailer driveshaft. The new arrangement gave a flat floor in the towing vehicle, where the transfer box had protruded above the floor in the Mk1 version, and also gave greater articulation. However, the Mk2 trailer could not be rolled through 360 degrees, as the Rubery Owen type could, without damage to the drivetrain or the risk of overturning the towing vehicle.

Land Rover agreed to give the Bushmaster Mk2 design a fair crack of the whip and built the relevant components into the first pre-production 961-series (RHD, Home Market, 24V) vehicle, which was registered as EXC 384L. The same system must have been built into the early production LHD 24V GS model (964-00008A) that went to Luxembourg for trials in 1975, because with it on those trails went a Scottorn Mk2 trailer, still carrying a number-plate reading EXC 384L. As explained in Chapter 7, the Royal Luxembourg Army decided not to buy any powered-axle trailers, but they did place a contract with Scottorn for undriven trailers that used 101 One-Tonne components.

POWERED-AXLE TRAILERS 155

Complication: the trailer-drive system brought with it an additional control lever in the cab. The one with the red knob engaged the centre differential lock on the towing vehicle; the green one engaged the winch drive; and the blue one engaged the trailer drive. AUTHOR

THE END OF THE AFFAIR

As explained earlier, the MoD was undecided about the value of the powered-axle trailer system as early as 1972, and one result was that only a small number of the pre-production 101s were built with it. Nevertheless, trials continued with the system at FVRDE. They were not without incident. One of the 101 prototypes fell victim to a weakness of the system that nobody appeared to have imagined. The unladen vehicle was towing a laden trailer up a 1-in-3 test slope when it lost traction. The trailer, however, did not lose traction, and as it pushed from behind, it hoisted the rear of the 101 into the air.

So there were already doubts about the system before the moment in 1974 that those associated with the 101 project remember as a critical turning-point. During a demonstration of Rover and Alvis military products for potential buyers, 04 SP 07 (chassis 961-00001A, also known as EXC 384L) was being used with a powered-axle trailer and towed gun. The vehicle got itself into a position where it was at 90 degrees to its trailer. As the driver applied power to drive off, the trailer gained traction before the towing vehicle did, and slowly pushed it over sideways.

George Mackie explained the occasion like this:

In order to get maximum cross-country performance, the [One-Tonne's] ground clearance was raised, which in turn meant raising the height of the chassis frame and therefore the trailer hitch point. One day, at a demonstration, the inevitable happened. As the outfit proceeded down a steep slope with a sharp turn at the bottom, the extra traction momentarily exerted by the trailer at right angles to the Land Rover had the effect of pushing it at the hitch point, which was rather high, and therefore turned it over!

The realization of this danger point, coupled with the very high cost of the Land Rover as a special vehicle only for the army, and the even higher relative costs of the trailer as a special trailer only for use with this particular Land Rover, got through to the book-keepers at the ministry, and the upshot was that the powered-axle trailer concept was abandoned.

As far as the MoD and the 101 One-Tonne were concerned, the idea was indeed abandoned. But that was not quite the end of the powered-trailer story.

POWERED-AXLE TRAILERS ■ 157

Problem, 1: an unforeseen weakness of the powered-trailer system was that if the trailer kept on going when the tractor lost grip, it could push the tractor over or into the air, as happened here on the 1-in-3 test hill at **FVRDE**.
TANK MUSEUM

Problem, 2: the military was already having second thoughts about the powered-axle system when an accident occurred at a demonstration. The unwieldy length of the Land Rover, limber and 105mm gun combination is also clear here. ROVER CO LTD

In 1971, T. T. Boughton & Sons Ltd (a forerunner of Reynolds Boughton) bought out the Scottorn trailer interests and turned the company into one of its subsidiaries on 31 January that year. Scottorn moved from New Malden in Surrey to Amersham in Buckinghamshire at the same time. Somewhat suppressed while the Rubery Owen design stole the limelight and while the new company settled down, the Scottorn design was resurrected in the mid-1970s and was further developed.

INDEX

3-litre engine 10, 11, 15, 18, 20, 29
105mm light jun 9, 16, 17, 20, 37, 73
110in Gun Tractor 10
956-series vehicles 67
957-series vehicles 68
959-series vehicles 68
961-series vehicles 70
964-series vehicles 70

additional order from MoD 59
ambulance 58, 75, 92
Andover aircraft 18
armoured 101s 98
assembly line 37, 53
Australian 101s 121
Australian trials 20, 21, 134
Avdelok rivets 58

BAOR 73
Barton, Tom 9
BDS conversions 70
Beeswing missile 30, 31
Belfitt, Cyril 9
Berlin Brigade 101s 79
Blindfire radar 70
British military 101s, quantities 76
Brunei 68, 70, 126
Busby, Norman 14, 15, 23, 24, 25, 26, 27, 29, 34, 53
buying a 101 144

Cammell Laird 111
Canadian trials vehicles 37, 39, 45, 71, 134
Challenger conversions 70, 99
changes from pre-production vehicles 53

chassis-cab models 53, 58, 91
chassis design 18
chassis numbers 63
chassis types 59
Chobham 9, 26
civilian 101s (plans) 47
CKD plans 53
colours 59
contract for engineering prototypes 19
contracts for production 101s 27, 74
conversions of 956-series vehicles 68
conversions of 959-series vehicles 70
conversions of 961-series vehicles 70
conversions of 964-series vehicles 71
Crathorne, Roger 45, 51, 52

deliveries (UK) 75
DND (Canada) 46, 47
dual-steer 101 101
Dubai 53, 70, 127

Egypt 60, 70, 71, 127
electronic repair workshop 101
end users of 956-series vehicles 68
end users of 957-series vehicles 68
end users of 959-series vehicles 70
end users of 961-series vehicles 70
end users of 964-series vehicles 71
ENV 11, 18, 23, 24

FACE vehicles 70, 102
Falklands War 16, 80, 94, 95, 99, 110
fleet code (trial vehicles) 40
Ford Falcon engine 21, 32

INDEX

forward-control designs, WD interest 12
FRT recovery vehicles 68, 102
funeral carriage 69
FVRDE 9, 10, 12, 15, 17, 20, 25, 26, 28, 40
FVRDE 'brochure' for 101 21
FVRDE forward-control mock-ups 12

Gama Goat 34
GS bodywork 58
GSR document 17
gun tractors 73, 103

hard-top cab 46
hood design 56
Hunt, Kevin 15

initial unit allocations 77
inspecting a used 101 145
Intercept Complex 71, 114

job cards 15
Judge Dredd 101s 138

Kenya 68, 128

last 101 built 61
Lees, Bob 15, 21
LETE (Canada) 46
Leykor 23
lightweight chassis 20
Llama (projected 101 replacement) 81
LT95 gearbox 23, 24
Luxembourg 53, 71, 128

Mackie, George 150, 156
Malayan interest in 101 28, 60, 134
Marshall's 32, 62, 70, 97
Martin-Hurst, William 23
Mayflower winch 20, 25
McWilliams, Bruce 23
Meakin, Terry 52
Milan missile carriers 70, 103
military asset codes (UK 101s) 90
military input to 101 design 14, 17
military serial numbers (UK 101s) 85
Miller, Geof 23
mortar carriers 70, 103

Mountbatten, Lord Luis 69
MVEE 26, 28

names used for the 101 71
Nokken winch 25, 55, 58, 62, 70

Oman 68, 132
one-ton vehicles in WD service 9
Operation *Corporate* 94, 95, 99
Operation *Wagon Train* 14
overseas sales 17, 53
owning a 101 today 143

paint schemes for UK 101s 86
'Pattern' GS bodies 62, 93
permanent four-wheel-drive system 23
Pittaway (commercial artist) 91
plastic fuel tank 34
Poole, Tony 50
powered-axle trailers 9, 37, 39, 56, 59, 73
pre-production models 37
pre-production models features 39
pre-production models, identities 40
pre-production models, list 42
production changes 62
production totals 67
prototypes of the 101 29

RAF 4 Wing 71
RAF 6 Wing 63, 68
RAF 101s 73, 89
Range Rover 22
Rapier tractors 75, 103, 110
Ras al Khaimah 70
reliability (REME report) 83
rework programmes 53, 62
Reynolds-Boughton RB44 81
ROF, Nottingham 68
Rover Company 9
Rover project team 15
Royal Marines 101 73
Royal Navy 101s 41, 73, 89
Rubery Owen 10, 151
running costs 145

Salisbury axles 23
Santana 2000 136

INDEX

Scottorn trailer drive system 11, 150
Scottorn Trailers Ltd 9
serial numbers of UK 101s 87
service life (UK 101s) 79
Seymour, Scott 15
Shaw, Frank 22
Shaw, John 15
Sheppard, Tom 56
Signals bodies 58, 75, 110
six-cylinder prototypes 18
six-stud axles 22, 23
Smith, AB 26, 34
South Africa 23, 134
Special Projects Department 50
Swingfire missile 30, 70, 91, 127

taper-leaf springs 25
technical specification, 6-cyl prototypes 36
technical specification, production models 71
Terra-Tires 28, 29
transfers between services 73, 89
Trans-Sahara Expedition 53, 55

trials, 1970 26
Twist, Ken 15, 27

UAE 70, 134
Uganda 68, 132
unknown end-users 134
unused vehicles 71
user trials (by MoD) 51

V8 engine 16, 17, 21, 23
V8 prototypes 23
Vampire 71, 116
vehicle identification plates 84
Vernon, Hugo 15
visions of the future 91
Volvo 17, 26, 27, 46

waterproofing 120
Wessex helicopter 17
winch-equipped 101s 92
winterization kit 41, 1